Also by Jamie Bartlett

Radicals: Outsiders Changing the World
The Dark Net
Orwell versus the Terrorists: Crypto-Wars and the Future of Surveillance

The People Vs Tech

How the internet is killing democracy (and how we save it)

JAMIE BARTLETT

3 5 7 9 10 8 6 4 2

Ebury Press, an imprint of Ebury Publishing
20 Vauxhall Bridge Road
London SW1V 2SA

Ebury Press is part of the Penguin Random House Group
of companies whose addresses can be found
at global.penguinrandomhouse.com

First published by Ebury Press in 2018

www.penguin.co.uk

A CIP catalogue record for this book is available
from the British Library

ISBN 9781785039065

Typeset in 12/15.5 Bell MT
by Integra Software Services Pvt. Ltd, Pondicherry

Printed and bound by Clays Ltd, St Ives plc

Contents

Introduction

IN THE COMING FEW years either tech will destroy democracy and the social order as we know it, or politics will stamp its authority over the digital world. It is becoming increasingly clear that technology is currently winning this battle, crushing a diminished and enfeebled opponent. This book is about why this is happening, and how we can still turn it around.

By 'technology' I do not mean *all* technology, of course. The word itself (like 'democracy') came from an amalgamation of two Greek words – *techne*, meaning 'skill' and *logos* meaning 'study' – and therefore encompasses practically everything in the modern world. I am not referring to the lathe, the power-loom, the motor car, the MRI scanner or the F16 fighter jet. I mean specifically the digital technologies associated with Silicon Valley – social media platforms, big data, mobile technology and artificial intelligence – that are increasingly dominating economic, political and social life.

It's clear that these technologies have, on balance, made us more informed, wealthier and, in some ways, happier. After all, technology tends to expand human capabilities, produce new opportunities, and increase productivity. But that doesn't necessarily mean that they're good for democracy. In exchange for the undeniable benefits of technological progress and greater personal freedom, we have allowed too many other fundamental components of a functioning political system to be undermined: control, parliamentary sovereignty, economic equality, civic society and an informed citizenry. And the tech revolution has only just got going. As I'll show, the coming years will see further dramatic improvements in digital technology. On the current trajectory, within a generation or two the contradictions between democracy and technology will exhaust themselves.

Strangely for an idea that nearly everyone claims to value, no one can agree on precisely what democracy means. The political theorist Bernard Crick once said its true meaning is 'stored up somewhere in heaven'. Broadly speaking, it is both a principle of how to govern ourselves, and a set of institutions which allow for sovereignty to be derived from the people. Exactly how this works changes from place to place and over time, but easily the most workable and popular version is *modern liberal representative democracy*. When I use the term 'democracy' from now on, this is what I'm referring to (and I am only looking at mature, Western democracies – to look beyond that is a different subject entirely). This form of democracy typically means that representatives of the people are elected to make decisions on their behalf, and that there is a set of interlocking institutions making the whole thing work. This includes periodic elections, a healthy civil society, certain individual rights, well-organised political parties, an effective bureaucracy and a free and vigilant media. Even that is not enough – democracies also need committed citizens who believe in the wider democratic ideals of distributed power, rights, compromise and informed debate. Every stable modern democracy shares nearly all of these features.

This is not another book-length whinge about rapacious capitalists who masquerade as cool tech guys, nor a morality tale about grasping multi-nationals. Democracy has seen off plenty of them over the years. While there are certainly contradictions in minimising tax while claiming to empower people, doing so doesn't necessarily betray insincerity. And, on first glance, technology is a boon to democracy. It certainly improves and extends the sphere of human freedom and offers access to new information and ideas. It gives previously unheard groups in society a platform and creates new ways to pool knowledge and coordinate action. These are aspects of a healthy democratic society too.

However, at a deep level, these two grand systems – technology and democracy – are locked in a bitter conflict. They are products of completely different eras and run according to different rules and principles. The machinery of democracy was built during a time of nation-states, hierarchies, deference and industrialised economies. The fundamental features of digital tech are at odds with this model: non-geographical, decentralised, data-driven, subject to network effects and exponential growth. Put simply: democracy wasn't designed for this. That's not really anyone's fault, not even Mark Zuckerberg's.

I'm hardly alone in thinking this, by the way. Many early digital pioneers saw how what they called 'cyberspace' was mismatched with the physical world, too. John Perry Barlow's oft-quoted 1996 *Declaration of the Independence of Cyberspace* sums up this tension rather well: 'Governments derive their just powers from the consent of the governed. You have neither solicited nor received ours. We did not invite you. You do not know us, nor do you know our world ... Your legal concepts of property, expression, identity, movement and context do not apply to us. They are all based on matter, and there is no matter here.' This is an exhilarating statement of the freedom offered by the internet that still holds digital aficionados in thrall. But democracy *is* based on matter, in addition to the legal concepts of property, expression, identity and movement. If you scratch beneath Silicon Valley's corporate pieties about connectivity, networks and global communities, you'll find that an anti-democratic impulse continues to exist.

In the following pages, I will argue that there are six key pillars that make democracy work, not just as an abstract idea, but also as a workable system of collective self-government that people believe in and support. These are:

ACTIVE CITIZENS: Alert, independent-minded citizens who are capable of making important moral judgements.

A SHARED CULTURE: A democratic culture which rests on a commonly agreed reality, a shared identity and a spirit of compromise.

FREE ELECTIONS: Elections that are free, fair and trusted.

STAKEHOLDER EQUALITY: Manageable levels of equality, including a sizeable middle class.

COMPETITIVE ECONOMY AND CIVIC FREEDOM: A competitive economy and an independent civil society.

TRUST IN AUTHORITY: A sovereign authority that can enforce the people's will, but remains trustworthy and accountable to them.

In the following chapters I will examine these pillars, and explain why and how they are threatened. In some cases they are under siege already. In other cases, I will look a little further ahead and argue that they soon will be. Whether it's the rise of smart machines limiting our capacity for moral judgement, the reappearance of tribal politics, or the prospect of mass unemployment as hyper-efficient robots displace break-taking humans, democracy is under assault from all sides. Some of these threats are familiar. There is nothing particularly new about angry politics, unemployment or citizen apathy, although they are taking a new form. But other threats are entirely novel: smart machines might replace human decision-makers, transforming political choices in ways we don't yet fully understand. Invisible algorithms are creating new, hard-to-see sources of power and injustice. As more of the world gets connected, it will be easier for a small number of rogue actors to cause immense damage and harm, often beyond the reach of the law. We don't have a clue how to deal with these problems.

In the final chapter I project how things might unfold if we continue on our current trajectory. We won't witness a repeat of the 1930s, everyone's favourite analogy. Rather, I believe that democracy will fail in new and unexpected ways. The looming dystopia to fear is a shell democracy run by smart machines and a new elite of 'progressive' but authoritarian technocrats. And the worst part is that lots of people will prefer this, since it will probably offer them more prosperity and security than what we have now.

But we shouldn't start smashing the machines just yet. For one thing, there is currently a tech arms race between democratic societies and their Russian and Chinese counterparts, and it is important for the democracies to win this race. And if subjected to democratic control, the tech revolution could transform our societies in myriad positive ways. However, both tech and democracy need to change dramatically. At the end of the book, I have 20 suggestions for how democracy – and more importantly, each of us – must change in order to survive in an era of ubiquitous intelligent machines, big data and a digital public sphere.

At this point you might well think I am a hypocrite, that I probably wrote this book on a laptop, used Google for my research, will tweet about the publication date and hope it sells strongly on Amazon. That's all true! Like many of us, I simultaneously rely on, love and detest all the technologies I write about. In fact, I have been working at the forefront of technology and politics for the last decade, at Demos, one of the UK's leading think tanks. Since I started there in 2008 I've written pamphlets about how digital technology would breathe new life into our desperately tired political system. Over the years my optimism drifted into realism, then morphed into nervousness. Now it is approaching mild panic. I still believe that technology can be a force for good in our politics – and that many of the big tech companies hope it can be, too – but for the first time I am genuinely concerned about the long-term prospects of the system that Winston Churchill once famously referred to as 'the worst kind of government, except for all the others that have been tried'.

The great tech pioneers, of course, do not share this concern because they are firm believers in a sunny techno-utopia and in their ability to take us there. I have been fortunate enough to interview some of them, and have spent a lot of time either in Silicon Valley itself or with people who inhabit that world. In my experience they are rarely evil and most have faith in the emancipatory power of digital technology. Many of the technologies they build are wonderful. But that makes them potentially more dangerous. Just like the eighteenth-century French revolutionaries, who believed they could construct a world based on abstract principles like equality, these latter-day utopians are busily dreaming up a society dictated by connectivity, networks, platforms and data. Democracy (and indeed the world) does not run like this – it is slow, deliberative and grounded in the physical. Democracy is analogue rather than digital. And any vision of the future that runs contrary to the reality of people's lives and wishes can only end in disaster.

Chapter 1: The New Panopticon

What the Power of Data is Doing to Our Free Will

We live in a giant advertising panopticon which keeps us addicted to devices; this system of data collection and prediction is merely the most recent iteration in a long history of efforts to control us; it is getting more advanced by the day, which has serious ramifications for potential manipulation, endless distraction and the slow diminishing of free choice and autonomy.

FOUNDING MYTHS ARE IMPORTANT for industries. They shape how companies see themselves and reflect how they wish to be seen by others. The founding myth for social media is that they are the heirs to the 'hacker culture' – Facebook's HQ address is 1, Hacker Way – which ties them to rule-breakers like 1980s phone phreaker Kevin Mitnick, the bureaucracy-hating computer lovers of the Homebrew Club scene and further back to maths geniuses like Alan Turing or Ada Lovelace.

But Google, Snapchat, Twitter, Instagram, Facebook and the rest have long ceased to be simply tech firms. They are also advertising companies. Around 90 per cent of Facebook and Google's revenue comes from selling adverts. The basis of practically the entire business of social media is the provision of free services in exchange for data, which the companies can then use to target us with adverts.*

This suggests a very different, and far less glamorous, lineage: a decades-long struggle by suited ad men and psychologists to uncover the mysteries of human decision-making and locate the 'buy!' button that lurks somewhere in our frontal lobe. A more cogent founding story is the early years of American psychology, which emerged as

---

* In 2017 $95 billion of Google's total revenue of $111 billion was from advertising, and it's a similar proportion for Facebook, Twitter and Snapchat. We sort of signed up to this of course, through the exchange: my data for your free service. You let people spy on you, and they give you an incredible service for free. But this exchange is a bit one-sided. Hardly anyone ever reads the terms and conditions they tick, because they're long and only comprehensible to a contract lawyer with a background in software engineering and one day a week to spare. A few years ago a British firm included a clause that asked for permission to 'claim, now and for evermore, your immortal soul', and no one noticed.

a serious academic discipline a century ago alongside the beginnings of mass consumer culture. Psychology had been developing in Europe – and especially Germany – for some years, and was imported to the US before the First World War. But the American variety diverged from the European fascination with philosophical whimsies like 'free will' and 'the mind'. Driven by pioneers such as James Cattell and Harlow Gale, it looked instead at how to turn the question of human decision-making into a hard science that could be used by business.[1]

In 1915 John Watson became president of the American Psychological Association. He argued that all human behaviour was essentially the product of measurable external stimuli, and could therefore be understood and controlled through study and experiment. This approach became known as behaviourism, and was later popularised further by the work of B.F. Skinner. The promise of malleable humans was catnip to companies hoping to sell products, and behaviourism spread through the corporate world like a virus. For some years, businesses – encouraged by Watson and others – believed they had godlike powers over desires, hopes, fears and, of course, shopping. Behaviourism was knocked out of fashion somewhat in the 1920s with the arrival of statistical market research (which, unlike behaviourism, actually required asking people questions). But together, behaviourism and market research signalled a more scientific approach to advertising that has been with us ever since.

If John Watson were alive today, he would be employed as 'chief nudger' at Google, Amazon or Facebook. Social media platforms are the latest iteration of the behaviourist desire to manage society through scientific observation of the mind, via a complete information loop: testing products on people, getting feedback and redesigning the model. Another word for this idea is what Yuval Noah Harari calls 'dataism': positing that mathematical laws of data apply to humans as well as machines. The notion that with enough data the mysteries of the human mind can be understood and influenced is perhaps the dominant philosophy in Silicon Valley today. In an oft-cited essay from 2008, then editor-in-chief of *Wired* Chris Anderson hailed the 'end of theory'. Scientific theories were unnecessary, he said, now that we have big data. 'Out with every theory of human behaviour ... Who knows why people do what they do? The point is they do it, and we can track and measure it with unprecedented fidelity.' Google engineers don't speculate and theorise about why people visit one site over another – they just try things and see what works.

In the bowels of every inspirationally branded tech firm some of the world's smartest minds are paid small fortunes to work out why you click on things, and to get you to click on more things. Although the secret of Facebook's success is ultimately the human psyche (humans are creatures that like to copy and watch each other and Facebook is the greatest system ever invented to allow us to see and be seen) this is supplemented by every imaginable tactic to keep you hooked. Nothing is left to chance, since even the smallest improvement can be worth a fortune. Tech companies run thousands of tests with millions of users – tweaking backgrounds, colours, images, tones, fonts and audio – all to maximise user experience and user clicks.[2] Facebook's homepage is carefully designed to be full of visible numbers – likes, friends, posts, interactions and new messages (and always in red! Urgent!). Autoplay, endless scroll and reverse chronological timelines are all sculpted to keep your attention.[3]

It's certainly working. Hordes of us are now members of a zombie army that walks while looking down at our phones and chats to distant disembodied avatars rather than whoever is sitting next to us. Like many people, I consider myself a witness to these changes rather than a participant, and so last year I downloaded an app called RealizD, which counted how often and for how long I checked my phone.

Monday 27th November: 103 pick-ups, 5 hours 40 minutes

Tuesday 28th November: 90 pick-ups, 4 hours 29 minutes

Wednesday 29th November: 63 pick-ups, 6 hours 1 minute

Thursday 30th November: 58 pick-ups, 3 hours 42 minutes

Friday 1st December: 71 pick-ups, 4 hours 12 minutes

According to these results, on average I pick up and check my phone 77 times per day. Take out sleep and that's roughly once every twelve minutes. I'm not alone. According to Adam Alter, addictions to alcoholism and tobacco are giving way to digital dependency, an epidemic of checking, picking-up, swiping and clicking.[4] Significant numbers of people now say they are addicted to the internet and could not live without their phone.[5] Some academics even think declining drug and alcohol intake among young people might be caused by them getting their dopamine rushes through pings and beeps.* 'In 2004 Facebook was fun,' writes Alter. 'In 2016 it's addictive.'[6] This is no accident. Welcome to the attention economy.

The reason I check my phone roughly once every twelve minutes is the constant but inconsistent feedback. Studies have shown that the anticipation of information is deeply involved with the brain's dopamine reward system, and that addictiveness is maximised when the rate of reward is most variable.[7] This is designed-in too, through the use of 'push notifications', which are the little beeps and messages that pop up to let you know when something has arrived in your inbox. Similarly, the introduction of a 'like' button in 2009 came from a much older subfield of – yes, this really exists – Liking Studies, which has long shown that likability is an advert's most potent characteristic.[8] (Apparently Facebook

---

* The popular explanation for this is that the beep or ping of a phone gives the brain a shot of addictive dopamine – the pleasure chemical. The truth is more sinister. Dopamine circuitry seems to predict how much we will enjoy something – and when it falls short, we feel a dopamine plunge. And so we keep at it, hoping to ramp it back up. The rush we feel of someone posting on our feed is the anticipation of what it will be. As you probably know from experience, the reality is never quite as good. We are stuck in an infinite loop of searching.

originally planned an 'awesome' button.)[9] Sean Parker, Facebook's first President, recently called the 'like' button 'a social-validation feedback loop ... exactly the kind of thing that a hacker like myself would come up with, because you're exploiting a vulnerability in human psychology'. He said that he, Mark Zuckerberg and others understood this, 'And we did it anyway'.[10]

Data

The Holy Grail for the social media giants, just as it always has been for all ad men, is to understand you better than you understand yourself. To predict what you will do, say and even think. Facebook doesn't collect data about you for fun; it does it to get inside your head. What the company knows about you, based solely on the untold hours you've spent there, is enough to fill several binders – interests, age, friends, job, activity and more. And that's not all. Facebook has partnerships with quietly powerful 'data brokers' like Acxiom, which has information on over 500 million active consumers worldwide, with thousands of data points per person: things like age, race, sex, weight, height, marital status, education level, politics, buying habits, health worries and holidays, often scooped up from other shops and records.[11] Armed with all this information, cross-referenced and analysed, companies can target you with ever more refined advertising.

Amazingly, this data collection frenzy is just getting started. By 2020 there will be around 50 billion internet-enabled devices – four times what there are now – and each one hoovering up data: cars, fridges, clothes, road signs and books. Your precious daughter playing with her doll: data point! Your partner adding some sugar to her tea: data point! Nothing will be safe from these giant, insatiable data monsters. Google has started to send Street View photographers into shops, offices and museums, in order to create detailed 3D models of the surroundings wherever you want to go. Smart homes want to know your preferred temperature, when you wash, what you cook, how long you sleep for. Everything will be collected, analysed and compared against everything else, in a relentless quest for dataism.

The data windfall is far beyond human analysis these days, which is why algorithms have become so central to the modern economy. An algorithm is a simple mathematical technique, a set of instructions that a computer follows in order to execute a command. That's the technical description, but in truth these are the magic keys to the kingdom, which filter, predict, correlate, target and learn. Your life is already guided by algorithms that determine everything from Amazon recommendations and your Facebook news feed to the things that pop up on your Google search. Your dating matches. Your route to work. Your music. News aggregators. Your clothes.

The scary thing about modern big data algorithms is how they can figure things out about us that we barely know ourselves. Humans are often quite predictable, and with enough data – even trivial or meaningless scraps such as what songs you play – algorithms can learn very important things about what sort of person you are.

Back in 2011, Dr Michal Kosinski, then a psychologist at Cambridge University, developed an online survey to measure respondents' personality traits. For decades psychologists have developed techniques to work out someone's personality through questionnaires.* Kosinski was interested in whether online data might determine something important

* The most common is informally known as 'the Big Five', which measures where people sit on five personality domains (Openness, Conscientiousness, Extraversion, Agreeableness and Neuroticism, or OCEAN for short) based on answers to standard questions, like how organised or disorganised you are.

about a person's personality without the need for a survey: perhaps it might be possible to generate a psychological profile simply based on things people had liked on Facebook. So Kosinski and his team set up several personality tests and posted them on Facebook, inviting people to respond. The surveys went viral – we do live in the age of narcissism, after all – and millions of people took part. By cross-referencing people's survey answers against their Facebook likes, he was able to work out the correlation between the two. From that he created an algorithm that could determine, from likes alone, intimate details of millions of other users who hadn't taken the survey. In 2013 he published the results, showing that easily accessible digital records of behaviour can be used to quickly and accurately predict sexual orientation, ethnicity, religious and political views, personality traits, intelligence, happiness, use of addictive substances, parental separation, age and gender.[12*]

---

* There is no evidence that Facebook uses these techniques. (Although, according to a May 2017 article published in the *Guardian,* Facebook's Australian team told advertisers they can detect when a teenager is stressed, insecure, deflated or anxious. Facebook responded that they do not 'offer tools to target people based on their emotional state'.)[13]

In 2017 I went to visit Michal Kosinksi at Stanford University, where he is now based. Many consider Stanford to be the university of Silicon Valley – it is nearby, and the founders of Cisco, Google, Hewlett-Packard and Yahoo! all graduated from there. Michal, who looks far too young to be a university professor, took me into his office in the Graduate School of Business (of course) and agreed to give me a demonstration of how this system works. I submitted my roughly 200 Facebook likes into his algorithm: *The Sopranos*, Kate Bush, *Terminator 2*, *The Spectator* magazine, etc. The algorithm went out into the world, looking at other people who had similar combinations, or variants of combinations. A little wheel span around on the screen for a few seconds while the algorithm worked its magic and the results popped out: open-minded, liberal, artistic and *extremely* intelligent. This is obviously a very accurate system, I told Michal. Far more bizarrely, it also determined that I was not religious, but that if I were, I'd be Catholic. I couldn't have put it better myself – I went to a Catholic comprehensive school aged 5–18, and while I have a soft spot for the religion, I am no practising churchgoer. Similarly, it predicted my job to be in journalism, and that I had a strong interest in history; I studied history at university and have a Master's degree in historical research methods.

All this from Facebook likes, which have nothing to do with my background or upbringing. 'This is one of the things people don't get about these predictions,' Michal told me. 'Obviously if you like Lady Gaga on Facebook, I can tell you like Lady Gaga … what's really world-changing about those algorithms is that they can take your music preferences or your book preferences and extract from this seemingly innocent information very accurate predictions about your religiosity, leadership potential, political views, personality and so on.' I'll show you in Chapter Three how political parties might use this at election time. But I left Michal's office with the sensation that this sort of insight was very exciting, but also a new source of power that we barely understand, let alone control.

The logical end goal of dataism is for each of us to be reduced to a unique, predictable and targetable data point. Anyone who's tried to talk to a chatbot or seen an ad for something they just bought knows that these technologies are far from perfect. But the direction of travel is clear, and it is easy to imagine the ways in which every choice you take might one day be subject to a series of algorithmically informed nudges, all carefully and perfectly calibrated around you. Just imagine! Get up nice and early, based on the auto-set alarm that knows your calendar and average getting ready time (factoring for typical traffic). A data-driven breakfast would be proposed after a quick analysis of the health stats of you and thousands of others like you, to ensure the perfect balance of nutrients you might need today. (Plus: a small reduction in your health insurance premiums, if you take its advice.) Hop in your driverless car, which is just returning from a night shift earning money for you as an autonomous taxi. And, as you relax into the journey, your personal AI assistant bot will advise you on what to say in today's key sales meeting, based on previous performance and who else will be present. Before being whisked back home …

The possibilities for advertising here would of course be phenomenal. If you fell off the diet bandwagon, or were even statistically likely to fall off it, based on an analysis of sleep pattern, diet, word use on Facebook and voice tone you would get an ad for the local gym. A personal AI assistant would be telling you things you need, exactly when you needed them, and you wouldn't even know why.

It's easy to lose sight of the positives, because this sounds like an episode of Charlie Brooker's *Black Mirror*. I run a centre at Demos that specialises in big data analysis, and we've found new ways to understand social trends, illness, terrorism and much more. Data can and will help people hold governments to account by making more information about departmental performance available. It's inevitable that we will one day have personal AIs that negotiate for us with company AIs (think credit cards, car loans, pensions and investments).[14] This is all good news from the user's perspective.

However, this whole pattern that leads from data collection to analysis to prediction to targeting presents three challenges to the life of a democratic citizen. The first is the question of whether being under the glare of social media and costant data collection allows people to mature politically. The second is the danger that these tools are used to manipulate, distract and influence us in ways that are not in our best interests. The third is more hypothetical and existential, concerning whether we even trust ourselves to make important moral decisions at all. We'll take each one in turn.

Life under the microscope

Back in 1890, in a landmark – and still highly relevant – article for the *Harvard Business Review*, Samuel Warren and Louis Brandeis (who would both later become Supreme Court Justices) questioned whether the arrival of the camera would put citizens at risk from constant surveillance. They realised that new technology often shifts delicate social norms, and therefore new laws are sometimes needed to keep up. The intensity and complexity of the early nineteenth century, thought Warren and Brandeis, meant that 'solitude and privacy have become more essential to the individual'. They argued that citizens needed a right 'to be let alone'.

Since then, the legal right of privacy has been enshrined in law and various measures put in place to protect citizens from both the overbearing state and unscrupulous companies, both of whom have reason to invade our private sphere. Without privacy laws – which vary greatly in force from country to country – we would today live in a world of total surveillance at all times. In countries where such laws don't exist, I fear it's almost certain that wearable tech, 'smart homes' and AI will create unprecedented levels of government surveillance and control.[15] This is not only a worry in oligarchies or autocracies. In free societies we're never 'let alone' either; the data gold rush has opened up new forms of potential surveillance from democratic governments, too, and most civil liberty groups worry what that means for legitimate political debate and activism. I, like many others, read with increasing alarm stories of people being arrested and prosecuted for saying things that are offensive and nasty, but no worse. In some cases intelligence agencies don't need to spy on you anymore; they can simply go to the technology companies and prise out of them what they need.*

There is another more subtle threat from Little Brother's constant surveillance and data sharing. Back in the eighteenth century, the philosopher Jeremy Bentham (of whom more later) proposed a new type of prison, which he called a 'panopticon'. It was designed so that all the inmates could potentially be observed by a single watchman – without any knowledge of when they were being watched. The possibility alone was enough, thought Bentham, to ensure that everyone behaved. Our modern panopticon doesn't have just one watchman: everyone is both watching and being watched. This kind of permanent visibility and monitoring is a way to enforce conformity and docility. Being always under surveillance and knowing that the things you say are collected and shared creates a soft but constant self-censorship. It might not feel like that when people are screaming abuse on Twitter – but for

* This, incidentally, is precisely what the NSA's PRISM programme did – which was the very first revelation made by whistleblower Edward Snowden.

every angry troll there are hundreds of quiet users, lurkers who watch but don't post, for fear of the angry Twitter mob, the data collectors, a nosy employer or the hordes of professional offence-takers who shark around the net waiting to be upset.

This is damaging to the citizen's ability to exercise moral judgement in their lives. Developing the faculties to think for oneself requires that people say controversial things, make mistakes and learn from them. But social media creates a strange form of performative politics, where we all act out certain roles and acceptable public responses (this idea is bad! This person is good!), which limits the room for genuine personal growth.[16] For example, the ability to forget is an important part of self-development, because changing one's mind is how we are able to mature and grow. As an increasing number of people – both famous and not – have found to their cost, digital technology never forgets. Sometimes that has the benefit of uncovering powerful people's motives and prejudices. But when one idiotic remark made on a forum when you were young and ill-informed exists forever, and can be dug up and republished exactly as it was, more and more people will conclude it is safer just to never say anything. This is not a good environment for the development of healthy, thinking adults.

The power of manipulation

The second problem is distraction and manipulation. We have always been subject to the influence of ads, branding and even the layout of supermarkets (sweets are always at children's eyeline). But the difference now is more than just one of scale. If data analysts or algorithms understand us better than we do, they can manipulate or control us in ways we can't understand or uncover.

Let's imagine a scenario. A personalised ad delivery system has learned to target you using language that it knows you would perceive as genuine and interesting, based on things you've said in the past. It crafts unique sponsored Tweets that push your emotional buttons, serves them up when you're usually online, and tailors them to whatever mood you're in at that moment. You had a bad-tempered encounter that day with a foreigner; you'll get a fearmongering anti-migration ad from the local politician. You did some recycling, too – and there's a reminder about solar energy from the local branch of Greenpeace.

These are relatively mundane things. But what if anti-Semites are targeted with increasingly personalised virulent content, simply because a model suggests they'll spend more time viewing it?

Or perhaps you can sell 20 per cent more anti-depressants to people if you catch them at a certain point in the week and use messages that play to low self-esteem. What if payday loan ads and gambling offers were precisely worded and targeted at the very moment that someone is most vulnerable or short on cash?

The extent to which our choices are ever truly free might depend on your view of free will (a group of philosophers known as 'hard determinists' question whether free will exists at all). But, at the very least, big data algorithms throw up important questions regarding new locations of power, influence and control. As a society we used to be very worried by hidden persuasion like this – so worried, in fact, that the US Federal Communications Commission once declared subliminal messaging to be 'contrary to the public interest' in spite of there being zero evidence of it ever influencing anyone. But no one even understands how modern manipulation works. I highly doubt any one employee fully understands all of a company's algorithms, much like no single worker could now make a car or a pencil. Powerful algorithms, comprised of lines upon lines of code, shape our world, and much of the time humans aren't involved at all. It's all automated and running on constant, self-improving feedback loops. And because it is intelligible to very few and kept under lock and key – the modern equivalent of the recipe for Coca-Cola – most regulators have no idea where injustice might occur, and much less how to keep on top of it.

Free as in freedom

With the arrival of ever more powerful artificial intelligence, the prospect of dataism – that belief that mathematical rules of data can be applied to human decision-making – may soon pose a fundamental challenge to our understanding of ourselves. This is the third and final problem. As Yuval Noah Harari explains in his remarkable book *Homo Deus*, for centuries we have believed that we are the ultimate source of meaning, and that human free will is the highest form of authority. That is our own founding myth: that we didn't conquer the world because we were stronger and bigger than other animals, but because we were smarter. We consider that democracy has a special moral value and purpose because we regard human judgement and moral choices as uniquely valuable.

But what if smart machines armed with petabytes of data were able to consistently make better, wiser and smarter decisions than us?

Futurists often talk about what this means for jobs (as I will, in Chapter Four) but rarely ask what it might do for our self-confidence and sense of worth. The speed of advance in artificial intelligence suggests that it will provide more and more practical insight and answers that are superior to humans'. Let's take medicine. Diagnosis by AI will outperform professional doctors within a few years (it already does in many areas, but regulation is slower than tech). At first we'll be nervous about trusting a machine with life and death decisions. But we will take to it quickly, just as we did with autopilot controls in planes. The same will be true of aid distribution, smart energy power grids, crop yield predictions, spotting oil leaks and more.[17]

It's only a matter of time before this shifts from practical questions to moral ones, since the distinction is not so fine as you might think. Ever since the French Revolution, each new wave of technology has been accompanied by speculation about how messy moral questions could be solved with pure, exacting science. It never has yet, but perhaps this time will be different, because of the intoxicating power of numbers. The most famous version of this dangerous idea came from our old friend Jeremy Bentham. The Panopticon wasn't his only idea; in 1789 he designed a 'felicific calculus', a kind of algorithm that would (he claimed) calculate the moral rightness of any decision, which he thought could be measured by whether it increased pleasure and reduced pain for the most number of people. His proto-algorithm took into account things like intensity, duration and fecundity. Bentham was a utilitarian because he thought the consequences of an action determined its moral worth. That's why he was enamoured by a machine. Consequences, unlike vague ethical theories like 'honour' or 'duty', can be measured and categorised.

Bentham didn't have either the computing power or the data to make his calculus really work. But imagine an algorithm that had access to zigabytes of data – people's moods, health, well-being, marital status, wealth, age and so on. Felicific Calculus 2.0 could model the likely outcomes of any decision on these variables – and work out some aggregate score to help you make a decision. Perhaps you could even programme your own moral convictions into the machine:

Me: I am a utilitarian. Should I buy organic food?

FC 2.0: Based on organic farming data, transportation emissions data, farming subsidy information, correlated against economic growth and projected national happiness data, the answer is 'yes' with a 62 per cent probability.

Me: I am a deontologist who believes in the categorical imperative. Can you verify that no one is harmed unnecessarily in the making of the latest trainers (and if not, please buy me a pair)?

FC 2.0: I have not ordered you a pair of the latest trainers. I advise you to consider changing brands.

And so on.*

Felicific Calculus 2.0 would never have enough data of course, and would be highly reductionist – a cold calculating machine that couldn't grasp the subtleties of human decision-making. This is what people said to Bentham at the time, too. But that wouldn't stop people leaning on it, because when it comes to making difficult decisions, humans (and government departments, in my experience) are easily swayed by numbers and data, no matter how imperfect. Numbers are intoxicating, because they

* Another view, even more technologically advanced and considerably more idealistic, would be to integrate, not to delegate machine morality. Instead of submitting our available choices of action to The Moral Judgement Machine and waiting for a response, integrate a morality module brain chip that becomes part of the mental decision-making process: the machine spits out the relevant possible consequences to varying levels of granularity, scored by utility and probability, and any other caveats.

hold out the promise of a pure, exact, judgement-free answer. Algorithms are doubly so, since they appear to be logical and objective calculating machines which draw on millions of examples.

Deny it if you want, but we already rely on the machine for moral choices. Cathy O'Neil, in her recent book *Weapons of Math Destruction*, documented dozens of instances where important decisions – relating to hiring policy, teacher evaluations, police officer distribution and more – are effectively outsourced to the cold and unquestionable efficiency of proprietary data and algorithms, even though these decisions have important moral dimensions and consequences.[18] I can imagine this kind of utilitarian thinking will take over the world, because it's amenable to data and AI. This would be a disaster. They might look and sound very objective but algorithms always start with a question framed by whoever is in charge. As a result they tend to reproduce the biases of their creators. That they are all currently created and owned by a bunch of rich white tech guys in Northern California has already thrown up some extremely unfortunate results. They are never neutral. For example, some police forces rely on data models to decide where to put police officers. However, crime tends to take place in poor neighbourhoods, which means more cops in those areas.

That generally means more people in those neighbourhoods getting arrested, which feeds back into the model, creating a self-perpetuating loop of growing inequality and algorithm-driven injustice.

Here's the real kicker. The danger is not in machines coughing up poor solutions, but the opposite. As they improve, they will repeatedly produce extremely good, money-saving solutions (at least, compared to human decisions), which will further establish their importance in our lives, even if they are unjust in invisible ways. If a machine diagnosis were repeatedly better than a human doctor, it would potentially be unethical to ignore the machine's advice. A government with a machine telling them a certain policing allocation would save money and cut crime would be hard to resist, even if it didn't solve any long-term problems.

Why wouldn't this extend to the very heart of the citizen's duty too? Already there is a proliferation of well-meaning apps designed to help you decide how to vote. You put in your views and preferences and the machine spits out a party for you. Nearly five million Brits have already used the voting app 'iSideWith' in multiple elections. The fact that five million people asked an app that they barely understood how to fulfil their most important duty as a citizen bothered exactly no one.

Before the 2015 UK general election, my think tank Demos helped design the methodology for a similar app called Verto. We all thought it was a brilliant idea at the time – I told everyone it would help voters understand where the political parties stood on different issues. I have now gone full circle, and believe they provide short-term convenience at the expense of undermining our long-term critical faculties. We should ditch them all.

If you're going to use an app, why not just hand your vote over to an algorithm entirely? Voters are notoriously bad at knowing even their own preferences. Classic democratic theory assumes an informed and attentive public, but in truth democracy is a highly inefficient way of arriving at decisions. We're far too irrational, and bring a headful of cognitive biases into the polling station. What if a vote-bot could scrape all your posting history, all your likes, all your friends' likes, and a thousand other metrics including salary, geography and family size, then look at all the alternative candidates and choose the candidate that most closely matches your interests. 'Siri – tell me how I should vote in the EU referendum.'

The possibility of AI smarter than us (by which I mean capable of repeatedly making better data-driven, practical decisions, rather than necessarily being more 'intelligent') would have profound implications for the nature of political and moral authority that we can barely imagine. After we've finished trying and failing to smash these machines, we might conclude – in small, harmless ways at first, of course – that perhaps we really don't know best and that it's probably wiser not to listen to ourselves when making important moral and political decisions. Perhaps eventually the world will be so complicated and confusing that a super-intelligent AI will be necessary just to keep it all moving along. What would be the moral case for humans to make significant decisions if there was another, superior, system? And what then would a democracy be, except an inefficient way to make consistently bad choices?

Futurists often talk about something they call the 'technological singularity'. It's the point at which machine self-improvement sparks a runaway, self-replicating cycle. Ray Kurzweil, *capo di capi* of futurists, undoubted genius and scientist at Google, has suggested this will occur around the middle of the century. (Others disagree.) Far sooner and more likely, in my view, is what I'll call the 'moral singularity' – the point at which we will start to delegate substantial moral and political reasoning to machines. This too would be a point of no return, for the same reasons: once we start relying on it, we'll never stop. The extent to which we believe ourselves to be moral agents depends on us repeatedly taking moral decisions – shopping, voting, raising children, joining campaigns and a thousand things besides. Yes, those choices are often riven with bias and error, which is why decision-by-machine will be so appealing. But our critical faculties only improve through the repeated use of reason, evidence and moral inquisition. This is no easy task, and requires that citizens are alert, aware, subject to a wide variety of challenging ideas and think carefully about the influences that bear on them. This is the duty of every citizen fortunate enough to live in a democracy. As algorithms get faster and smarter, the pressure will grow to hand these inconveniences off for the sake of ease, speed or ignorance. To give in to this urge will be easy, but it would condemn us to losing the ability to think freely, and to spiral into ever greater reliance on machines. Given how bad we often are at making difficult decisions, the result might be a wiser and more humane society. But it would be difficult to call such a place a democracy.

Chapter 2: The Global Village

Why the Closer We Get, the Further We are Apart

Information overload and connectivity has encouraged a divisive form of emotional tribal politics, in which loyalty to the group and anger outrank reason and compromise. While partisanship is necessary in politics, too much of it is dangerous. Political leaders are evolving to the new medium of information – hence the rise of populists who promise emotional, immediate and total answers. But warring tribes of anchorless, confused citizens is a precursor to totalitarianism.

BACK IN THE 1960s, celebrity academic and cryptic cultural theorist Marshall McLuhan predicted that the coming age of electronic communications would lead to the breakdown of established structures and identities. The consequence, he asserted, would be a return to a more tribal society. He famously called this seamless web of information 'the global village'.[1] People at the time celebrated this idea.

McLuhan remains a distant inspiration for Silicon Valley, one of the original thought leaders and intellectual rock stars of the tech revolution. His 'global village' still bounces around Palo Alto, Mountain View and Cupertino. Every time you hear talk of 'global communities' and 'total connectivity', it's the ghost of McLuhan. 'By enabling people from diverse backgrounds to easily connect and share their ideas,' wrote Mark Zuckerberg back in the early days of his site, 'we can decrease world conflict in the short-term and the long-term.'

McLuhan, the great prophet, was far too smart not to hedge his bets. He also said that conflict and disharmony was possible in a world where everyone was connected to everyone else, because information-at-all-times would be so discombobulating that it would spark a mass identity crisis. 'The day of political democracy as we know it today is finished,' McLuhan told *Playboy Magazine* in a 1969 interview. 'As man is tribally metamorphosed by the electric media, we all become Chicken Littles, scurrying around frantically in search of our former identities, and in the process unleash tremendous violence.'[2] CEOs, high-profile endorsers, hangers-on, early technologists and politicians all tended to ignore this bit, because these sorts of people much prefer optimism to tremendous violence.

Democratic politics has always been raucous and often extremely divisive. Think the Iraq or Vietnam Wars, or even the miners' strikes of the 1980s. But since the Second World War, democratic politics has been, by and large, marked by remarkable levels of civility and shared assumptions, even if there were occasional flashpoints of fury. Over the last few years however, the nature of political disagreement has changed. It's gone tribal. It is becoming hyper-partisan, characterised by fierce group loyalty that sometimes approaches leader worship, a tendency to overlook one's own failings while exaggerating one's enemies and a dislike of compromise with opponents. Politics is becoming like sport. Supporters of Corbyn and Trump would detest the comparison, but note how both have the merchandise, the shouting supporters, the victory chants and the hagiographies. We are living, as McLuhan predicted, through a great re-tribalisation of politics. I say *re*-tribalisation, because tribal loyalties and identity have characterised human existence for far longer than modern politics. We've had enough civil wars to know that the need to belong to a group is deep-rooted.

Psychologists have long remarked on the irrationality of large crowds. Charles Mackay, (before the more famous work of Gustave Le Bon), wrote that '[men] think in herds ... it will be seen they go mad in herds, while they only recover their senses slowly, one by one'. The early proponents of representative democracy – especially the US Founding Fathers – feared the inflamed passion of the mad mob. They looked back through history and how 'the common impulse of passion or of interest, adverse to the rights of other citizens' was the 'moral disease under which popular governments have everywhere perished'.[3] They were partly scared of poor people – a feeling not unusual among the ruling classes – but they were also wise old heads who knew that democracy carries the risk of degeneration into ignorant and self-interested mob rule, running on unchecked emotion. They carefully designed a representative system with checks, balances and periodic elections that would delegate authority upwards and serve as a check on the confusion and fury of the multitude.

Anyone who has spent more than five minutes on Twitter would recognise Charles Mackay's definition. The modern technologist, however, doesn't believe the mob is faceless and mad, but rather wise and just: he reads books like Howard Rheingold's *Smart Mobs* and talks endlessly about 'crowd sourcing' solutions and 'the wisdom of the crowd'. He trusts the 'hive mind'. Crowds certainly are wise when it comes to solving technical, non-value based problems like fixing computer bugs, but politics is very different.[4]

Humans were perfectly good at killing each other because of politics long before the iPhone turned up. But Silicon Valley, in its optimistic quest for a global village of total information and connectivity, has inadvertently let tribalism back out of the cage that modern representative democracy built for it.

The great clustering

One of the most important – and sudden – changes in politics for several decades has been the move from a world of information scarcity to one of overload. Available information is now far beyond the ability of even the most ordered brain to categorise into any organising principle, sense or hierarchy. We live in an era of fragmentation, with overwhelming information options.

The basics of what this is doing to politics is now fairly well-trodden stuff: the splintering of established mainstream news and a surge of misinformation allows people to personalise their sources in ways that play to their pre-existing biases.[5] Faced with infinite connection, we find the like-minded people and ideas, and huddle together. Brand new phrases have entered the lexicon to describe all this: filter bubbles, echo chambers and fake news. It's no coincidence that 'post-truth' was the word of the year in 2016.

At times 'post-truth' has become a convenient way to explain complicated events with a simple single phrase. In some circles it has become a slightly patronising new orthodoxy to say that stupid proles have been duped by misinformation on the internet into voting for things like Brexit or Trump. In fact, well-educated people are in my experience even more subject to these irrationalities because they usually have an unduly high regard for their own powers of reason and decision-making.*

What's happening to political identity as a result of the internet is far more profound than this vote or that one. It transcends political parties and is more significant than echo chambers or fake news. Digital communication is changing the very nature of how we engage with political ideas and how we understand ourselves as political actors. Just as Netflix and YouTube replaced traditional mass-audience

---

* According to one recent book on partisanship, people with the strongest interest in and knowledge of politics are the most likely to selectively interpret information to suit their own biases.

television with an increasingly personalised choice, so total connection and information overload offers up an infinite array of possible political options. The result is a fragmentation of singular, stable identities – like membership of a political party – and its replacement by ever-smaller units of like-minded people.[6] Online, anyone can find any type of community they wish (or invent their own), and with it, thousands of like-minded people with whom they can mobilise. Anyone who is upset can now automatically, sometimes algorithmically, find other people that are similarly upset. Sociologists call this 'homophily', political theorists call it 'identity politics' and common wisdom says 'birds of a feather flock together'. I'm calling it re-tribalisation. There is a very natural and well-documented tendency for humans to flock together – but the key thing is that the more possible connections, the greater the opportunities to cluster with ever more refined and precise groups. Recent political tribes include Corbyn-linked Momentum, Black Lives Matter, the alt-right, the EDL, Antifa, radical veganism and #feelthebern. I am not suggesting these groups are morally equivalent, that they don't have a point or that they are incapable of thoughtful debate – simply that they are *tribal.*

What transforms a group of like-minded people into a motivated, mobilised tribe is a sense of shared struggle and common grievance. And the internet is the largest and most abundantly stocked pantry of grievance in the history of mankind.

If you are a transgender person, you can cite and share the awful crime statistics regarding violence towards trans people.

If you are a person of colour, survey data reveals over and over the enormous differences in your life opportunities.

If you are white and working class, studies find that your group has the lowest likelihood of getting to university, and the lowest sense of personal agency.

If you are a Muslim, you're more likely to end up in prison.

If you are middle class, academic research has found the last 30 years of globalisation has resulted in an unprecedented decline in your wages.

If you are a woman, you are still earning less than men, and are subjected to a huge amount of casual sexism.

If you are a man, you are subject to reverse discrimination, will live for less time and are significantly more likely to die from suicide.

I don't mean to denigrate these issues, since all the above statements are true and reflect genuine problems. The point is that every individual now has a truckload of reasons to feel legitimately aggrieved, outraged, oppressed or threatened, even if their own life is going just fine. For some people, being generally decent, this produces a powerful sense of belonging and solidarity with a group they never even thought about until they kept reading how oppressed they were.*

For years we've retreated into tribes with different information and principles, but the internet has opened new ways of forming, finding and joining ever-smaller tribes that we never even knew we belonged to, and stuffing ourselves full of evidence to harden the conviction. I see this happen all the time. I'm a white man in his late thirties and went to a comprehensive school. This is pretty unremarkable. But the more I read online how white working-class boys in comprehensive schools are the worst performers, suffer from high levels of suicide and so on, the more I identify

---

* It surprises me that so few writers remark on the relationship between tribal or identity politics and the internet. Lots of people write angrily about identity politics, and yet rarely ask whether the internet is encouraging it. (My own theory is that many critics of identity politics are libertarian free-speech activists and so are reluctant to blame the internet, because of how it undeniably helps freedom of expression.)

as a member of that tribe. Tribalism is understandable, but ultimately it is damaging to democracy, because it has the effect of magnifying the small differences between us, and transforming them into enormous, unsurpassable gulfs.

The 'system one' system

McLuhan had a theory that the written word, and by consequence the literate man, was calm, cool and rational. He ordered things and categorised them, which meant he had time to carefully analyse them. He was a reflection of the medium through which he received information. (This is where his other famous saying 'the medium is the message' comes from.) By contrast he thought electronic information, and especially television – his era's internet – was aural. It was sound and pictures, a more complete sensory experience. If literate man was rational, said McLuhan, then electronic man would be more emotional, aural and tactile.

McLuhan's prescient 50-year-old 'probes' (he called his ideas *probes*) into how technology would change behaviour are still significantly more insightful than almost every 'thought-provoking' TED Talk. But McLuhan wasn't a scientist. He didn't conduct studies or test theories. Fortunately Daniel Kahneman, the academic most associated with examining bias in human decision-making, did. Through decades of empirical research with long-time collaborator Amos Tversky, he pioneered the study of how we take decisions – and especially irrational ones. I won't recite the Stanford Prison Experiments or the Ultimatum Game, but Kahneman's main point was that there are two basic systems that govern human behaviour. 'System one' thinking is fast, instinctive and emotional. It's the reptilian brain, running on instinct. By contrast, 'system two' thinking is slow, deliberative and more logical.[7] It sometimes, but not always, acts as a check on those wilder rages.

Modern democracies aspire to run on 'system two' logic, and its ideal citizens are McLuhan's literate man. Its institutions are arranged to arrive at logical, thought-out, fact-driven decisions. The internet, by contrast, more closely resembles 'system one': everyone and everything is immediate, instinctive and emotional.

The internet inculcates new assumptions about how things should work in two important respects. First of all, everything online is fast and personalised: access to everything and everyone, to millions of web pages, all the goals, all the baby pictures, and all for free. You zoom in, you zoom out, swipe, tap and chat with a far-flung relative. As Douglas Rushkoff explains in his recent book *Present Shock*, in the modern world 'what we are doing at any given moment becomes all-important'. Consider it! You once queued to pick up (and pay for) photographs that you took a week ago, without knowing if they were even any good. The result is a growing disconnection between the choice and freedom that characterise our lives as consumers, and the compromise and tedious plodding world of politics. Note how, for example, so many people who disagree with Brexit use the language of a small child that has yet to develop a theory of mind: why should *I* accept the result, *I* didn't vote for it and *I* want *my* country back.*

* Here's a prediction: any new 'populist' party formed in the coming years will promise to introduce more referendums and digital voting for members. They will say that it's ordinary people speaking out against the establishment, and hail it as a new way to break the slow, corrupt 'system'. (They will obviously ignore whatever the people say if they don't like it.) In fact, the inevitable consequence of digital technology generally will be an increased demand for more referendums. The very real prospect of secure, affordable e-votes opens the possibilities of public votes every week on every subject. This is a highly alluring trap, which will only further spur the rise of 'system one' politics. If the Brexit and Scottish independence votes showed anything, it's that single issue plebiscites: a) don't actually 'settle' questions and b) are extremely divisive because they force people into binaries rather than seeking compromise.

Second of all, the internet is primarily an emotional medium, which is something that many technologists fail to grasp. Speed and emotion are related, of course, because both are means by which our finite brain handles information overload and total connectivity. It is obviously true that citizens need information to form opinions and make judgements, and there are many benefits to a more democratic form of media. But the modern citizen is expected to sift through an insane torrent of competing facts, networks, friend requests, claims, blogs, data, propaganda, misinformation, investigative journalism, charts, different charts, commentary and reportage. This is confusing and stressful, and so we lean on easy and simple emotional heuristics to make sense of the noise. As has been well documented, we rely on 'confirmation bias' – reading things we already agree with, surrounding ourselves with like-minded people and avoiding information that does not conform to our pre-existing view of the world. Similarly, because there is so much noise out there, studies repeatedly find that emotional content is more likely to get traction online – shares, retweets, etc. – than serious and thoughtful comment and stories. For example in the final months of the 2016 US election, fake election news stories on Facebook – always misleading, emotional, angry, outrageous and wrong – were shared more widely than the more sober analysis from the *New York Times* or *Washington Post.*[8]

The internet has not created this problem. We've always had our affiliations and always been roused by emotion. Liberals have always read the *Guardian* or the *New York Times,* and conservatives the *Telegraph* or the *Wall Street Journal.* But the internet has taken this to a whole new level. And just wait. Within a couple of years, video manipulation will be extremely believable and widely available. Anyone will be able to make any public figure 'say' anything they want, making it indistinguishable from the real thing. Fake videos could circulate of Donald Trump saying he's secretly a member of the Ku Klux Klan, or that George Soros is funding an anti-democratic coup.

The problem with tribes

Tribalism and 'system one' thinking are the direct products of information overload. These are ideal conditions for division and disagreement to turn into existential opposition. There's nothing wrong per se with political tribes. In democracies some degree of partisanship is necessary and even desirable.[9] But if partisanship overwhelms everything democracy breaks down because it makes compromise impossible. Reason and argument give way to emotion and blind tribal loyalty.

Understanding this process – of how opponents become enemies – is one of the most important questions facing modern democracies. Tommy Robinson, former leader of the English Defence League, offers a good example of how this transformation happens, and the role the internet plays in bringing it about. For several weeks during 2015–2016 I followed Tommy across Europe. Whenever I was with him, he'd regularly scan Twitter for grim stories he could share with supporters. On Tommy's Twitter feed on 17 December 2017, a day I picked at random, he shared stories about: gay men being attacked by Muslims and told they are not welcome in Walthamstow (*Evening Standard*); Sikhs being told to convert to Islam by Pakistani officials (*Rabwah Times*); an Italian town removing a Christmas tree (*Voice of Europe*); armed police patrolling the town centre in Luton amid a massive terror threat (*Westmonster*); a Somali refugee claiming benefits from the UK while living in Somalia (*Mirror*); a former anti-terror chief predicting a terror attack before Christmas (*Daily Mail*); twin suicide bombers attacking a church in Quetta Balochistan (*Reuters*); and a continuing terror threat forcing King's College

Chapel to scrap the traditional queue for its Christmas carol concert (*Telegraph*).

Often, the stories that Tommy shares are not made up. A decent number come from respectable, mainstream news outlets and are reporting things that really are happening. The ability to so tailor one's news consumption – reading the same issues over and over and then firing them out to thousands of others – has a powerful effect. According to Joel Busher, an academic who spent sixteen months embedded with the EDL, supporters use certain 'frames' to make sense of all this, tagging particular phrases onto each anecdote: the 'incompatibility of West and Islam', 'cultural Marxism' controlling public life, a 'two-tiered' system set against the white British. That transforms one-off stories into explanations of the perceived injustices they face, and is used to unlock meaning and stimulate emotional responses.[10] Any positive stories about Islam, those that can balance or lend perspective to isolated incidents, are simply washed away in a sea of negativity, or framed as propaganda or liberal journalists refusing to accept the truth.[11] Tommy has spent so long reading the same thing that opponents stop being merely people with respectable differences of opinion – how could they be when the problems and answers are apparently so obvious?

With so many seemingly evident problems and answers available, opponents can only be incoherent babblers, sinister Machiavellians, oppressors who don't understand your suffering. Disagreement over practical issues starts to involve purity and impurity: at which point there are no negotiable principles, just team loyalties. 'We' are good and pure, while 'they' are evil and corrupt.[12] There are signs of this divisive polarisation everywhere. Not just the fact that people have different political views, but that different views are a sign of deeper moral defects. According to the polling company YouGov, three quarters of young people who voted Remain think old people are prejudiced, and a similar proportion of older Leave voters believe young people are entitled and unwilling to work hard. In the US 'very unfavourable' views of supporters of the opposing political party more than doubled between 1992 and 2014, and by 2016 around half of Americans believed the opposition (but not them) were 'closed-minded'.[13] Once-serious newspapers now write headlines in which judges are 'enemies of the people' and principled MPs are 'saboteurs'. It is of course a vicious circle.

Have you ever noticed how swiftly arguments online seem to progress from mild disagreement to absolute denouncement? In my experience, people who voted Leave and Remain in the 2016 EU referendum will generally get on amicably enough around the dinner table. They will disagree of course, but they will at least listen and attempt to understand each other. Online, however, Remain voters are recast as smug and elitist 'remoaners', while Brexiteers are irresponsible nativists and hate-filled jingoists. Once more, the internet's mode of communication is driving this problem. The liberals' hopeful theory about the role of debate is that coming into contact with opposing views and opinions can help resolve difference. However, decades of studies have found that getting someone to change their mind about anything is extremely difficult. 'Beliefs are like fast cars,' writes the neuroscientist Tali Sharot. 'They affect our well-being and happiness … we try to fill our minds with information that makes us feel strong and right, and to avoid information that makes us confused or insecure.' This is why, when exposed to contradictory facts, most of us become *more* strongly set in our beliefs. Note, for example, how much Donald Trump was negatively fact-checked, and how little difference it made to the result of the election.[14] Several inconvenient studies have found that if two groups of people debate with each other they often consequently hold more extreme views than when they started.[15] No one knows why exactly (some studies say that it is an evolved trait that helps us cooperate, a kind of 'my-side bias'). Under certain conditions, we can and do change our minds, of course. Arguments that are carefully made, detailed, and involve an understanding of other people's mindset or background can make a difference.[16] But it's usually a slow and laborious process.

Unfortunately, the nature of digital communication rarely allows for that, because interaction with rivals or opponents is typically swift, fleeting and emotionally heightened. It therefore does little to improve understanding – and can often do the opposite. In 2001, cyber-psychologist John Suler explained why this was, listing several factors that allow users of the internet to ignore the social rules and norms at play while offline. Suler argues that because we don't know or see the people we are speaking to (and they don't know or see us), because communication is instant, seemingly without rules or accountability, and because it all takes place in what feels like an alternative reality, we do things we wouldn't in real life. Suler calls this 'toxic disinhibition'. This is what all the articles about 'echo chambers' and 'filter bubbles' miss. The internet doesn't only create small tribes: it also gives easy access to enemy tribes. I see opposing views to mine online all the time; they rarely change my mind, and more often simply confirm my belief that I am the only sane person in a sea of internet idiots.

It's unfair to lay all this at the door of Big Tech, since much of this is a human, not technological, weakness. Tech has turbocharged these weaknesses, but we are also responsible. And let's not romanticise life before the internet. People have always clustered, and politics has always been divisive. There have always been manipulation and liars in politics, too (Harry Truman once said of Richard Nixon that he was a 'no-good lying bastard, who can lie out of both sides of his mouth, and if he ever caught himself telling the truth, he'd lie just to keep his hand in').

However, the tech giants have turned these psychological weaknesses into a structural feature of news consumption and exploited them for money. Their financial incentives sometimes run directly contrary to the democratic need for people to be informed and draw from a wide range of accurate sources and ideas. All social media platforms insist they are 'platforms' not 'publishers' which means that, unlike newspapers, they aren't legally liable for the content they host. This protection (known as the 'mere conduit' clause under EU law) is extremely important for companies like Facebook or YouTube, because without it they would have to somehow check the billions of pieces of content uploaded to their sites. As a result they are hesitant to intervene too much in cleaning out divisive or misleading media, in case lawmakers conclude they are behaving like publishers and regulate them as such. They have no obvious escape from this dilemma.

But being apparently neutral is itself a kind of editorial decision. Everything on social media *is* still curated, usually by some mysterious algorithm rather than a human editor. These algorithms are designed to serve you content that you're likely to click on, as that means the potential to sell more advertising alongside it. For example, YouTube's 'up next' videos are statistically selected based on an unbelievably sophisticated analysis of what is most likely to keep a person hooked in. According to Guillaume Chaslot, an AI specialist who worked on the recommendation engine for YouTube, the algorithms aren't there to optimise what is truthful or honest – but to optimise watch-time. 'Everything else was considered a distraction,' he recently told the *Guardian*.[17]

These non-decision decisions have huge implications, because even mild confirmation bias can set off a cycle of self-perpetuation. Let's say you've clicked on a link about left-wing politics. An algorithm interprets this as you expressing an interest in left-wing politics, and therefore shows you more of it. You're more likely to click again, since that's the choice in front of you – which is interpreted as another signal. According to research conducted by Chaslot since he left YouTube, the company systematically amplifies videos that are divisive, although YouTube denies this. To be fair, no one I've ever met in tech is happy about this: over the last couple of years most have recognised that it is an issue and have promised to fix it. But the problem is that no one is intentionally programming it to be sensationalistic – it's just a mathematical response to our general preference for edgy and outrageous videos. This is both a mirror and a multiplier: a giant feedback loop powered by big data. You feed data in, and you get results that replicate themselves. Newspapers have always traded on outrage and sensationalism, because they've long known what algorithms have recently discovered about our predilections. However, the difference is that newspapers are legally responsible for what they print, and citizens generally understand the editorial positions of various outlets. Algorithms, however, give the impression of being neutral and can't be held to account – even though the YouTube algorithm alone shapes what 1.5 billion users are likely to see, which is more than every newspaper in the world combined.

Trump the tribal

September 26, 1960, changed politics forever. That evening, in the hotly anticipated presidential debate, the relatively unknown Senator John Kennedy squared off against Vice President Richard Nixon. Those listening on the radio thought that Nixon had won. But this was the first time a presidential debate was televised – and by 1960, 88 per cent of Americans had a TV, compared to just 10 per cent a decade before. Unlike the radio listeners, the millions who watched on TV thought the fresh-faced Kennedy had trounced the sweaty, pallid Nixon. By the next morning, Kennedy was a star – and of course he went on to win the election. From that point on, telegenicism was regarded as necessary for any political hopeful.

Political leaders evolve, albeit slowly and imperfectly, to fit the medium through which they reach the people. The latest iteration, and the first bona fide politician of the social media age, is Twitter addict and world-class simplifier Donald Trump. He is the leading act in a new cast of populists who have found the internet a revelation for their style of politics, including Nigel Farage, Bernie Sanders, Dutch anti-Islam firebrand Geert Wilders, Italian comedian and Five Star Movement founder Beppe Grillo and others. Some of them are left wing, some are right wing, but all are 'system one' leaders who became popular by promising easy solutions to complicated questions. Trump is the strong man, the tribal leader who trades on outrage. He offers swift, immediate and total answers: it's the fault of the bureaucrats, the politically correct media, judges and immigrants. He promises to deliver the people quickly and completely from the complexities of the world. And above all, he offers a sense of tribal belonging in a digital world characterised by confusion, uncertainty and information overload. He represents all the problems described in the last few pages, in human form. It was perhaps inevitable that growing inequality and globalisation would result in some form of political backlash. But the form it has taken is a reflection of our media.

In her masterpiece *The Origins of Totalitarianism* Hannah Arendt warned that if citizens float around like corks in a stormy sea, unsure of what to believe or trust, they will be susceptible to the charms of demagogues. When she wrote *Origins* in the 1950s, she never imagined the digital world, but she would have recognised how an anchorless and bewildered post-identity mob would demand a tribal leader who could bring order to chaos, simplicity to complexity and a sense of belonging.

If you listen carefully to many of Trump's supporters there is a palpable sense that he – The Donald – is a leader who has come to save them from threats to the tribe from all sides: liberals, Muslims, Mexicans and the mainstream media. It is worth considering whether walls, travel bans and extreme vetting are not simply part of the urge to keep enemies of the tribe out. Even his opponents are secretly obsessed with him, and I suspect many of them admire his gall. 'Everyone's sticking together in their groups,' a Long Island housewife told the *New York Times* in mid-2016, 'so white people have to, too.'[18] This is tribal politics, but applied on a mass scale. A blind loyalty to a man, a faith in his powers, indifference to truth and a desire that certain groups should stick together. Indeed, several studies conducted after the election found a strong correlation between the importance of white identity and support for Trump. Richard Spencer – an influential alt-right leader – summed it up like this: 'As long as whites continue to avoid and deny their own racial identity, at a time when almost every other racial and ethnic category is rediscovering and asserting its own, whites will have no chance to resist their dispossession.'[19]

If the medium is the message, is there a way to escape the drift toward ever more extreme 'system one' tribal politics? Of course. Laws, regulations or education can help. But, in the end, a cultural shift in how we understand political difference is the only real solution. When he became president in 1861, Abraham Lincoln appointed his opponents in the Republican primary race to prominent positions within his cabinet. He recognised that his rivals were men of great talent too, and that having access to a wide range of opinions would sharpen his own thinking. As Doris Goodwin, author of *Team of Rivals*, explains, it was often Lincoln's leadership qualities that held the group together: 'He understood that human relations are at the core of politics and that if you deal with people in the right manner, you are going to be able to work effectively with them. The qualities we associate with human greatness – such as sensitivity, empathy, compassion, kindness, and honesty – are also keys to political success. When something went well, Lincoln always shared the credit. When something went wrong, he shouldered his share of the blame. When he himself made a mistake, he acknowledged it immediately.

He made time for each of his cabinet members, so that they all felt they had access to him. He treated them all respectfully and fairly.' This approach – of recognising one's own weaknesses and one's opponents' strengths, and working together on the basis of common human decency – can keep tribalism in check. We can't wait for another Lincoln to turn up, but each of us can apply this approach in our political lives.

But it will take work, time and effort. Otherwise there's no knowing what depths politics can plumb, but a system comprising warring and emotional tribes, surrounded by grotesque enemies and a gutful of malleable incendiary 'facts' doesn't lead anywhere good. When compromise and negotiation break down, in the end only coercion and violence can resolve differences of opinion, and this is where we are heading. On 16 June 2016, Labour MP Jo Cox was murdered by Thomas Mair, a neo-Nazi who called her a 'collaborator' and a 'traitor' due to her campaigning in favour of the EU. Even this tragic event did not temper the rage. In late 2017 the Committee on Standards in Public Life reported on the intimidation parliamentary candidates had been experiencing: 'persistent, vile and shocking abuse, threatened violence including sexual violence and damage to property'. Over half of candidates surveyed were fearful about abuse and intimidation – and every single female MP active on Twitter had received abuse.[20]

At a campaign rally in Iowa in January 2016, Trump told his supporters that he could 'stand in the middle of Fifth Avenue and shoot somebody and … wouldn't lose voters'. There is a distinct and terrifying possibility that, in an era in which emotion outranks truth, bias outranks objectivity and tribe outranks compromise, he was right.

Chapter 3: Software Wars

How Digital Analytics Has Changed Elections

Donald Trump's digital campaigning in the 2016 presidential election showed how big data and micro-targeting can win votes. The continuing evolution of these digital techniques will change the type and style of politicians we elect – and more importantly, it will mean more power for rich groups to influence elections in ways we don't understand.

ONE SUNDAY AFTERNOON IN May 2016, Theresa Hong, a digital communications specialist with several years' experience in political campaigns, was at home in San Antonio, Texas, when her phone pinged.

'Theresa – this is Brad Parscale. Are you able to write anything?'

Theresa knew Brad pretty well – like her, he orbited the city's PR scene. A moderately successful tech entrepreneur from Kansas, Brad had lived in San Antonio since graduating from university in the late 1990s. In 2010, after a few years hustling a living with various digital businesses, he was asked to build a website for Donald Trump's real estate division and impressed his employer with his loyalty and hard work.[1] When Trump declared his bid for the Republican nomination Brad was drafted in to run the digital campaign. Although the Republican Convention wasn't until June 2016, by late April it was increasingly clear that Trump would be the nominee, and Brad was well-placed to work on the presidential campaign, too.

Brad and Theresa share more than just a profession and right-leaning politics. Both are in their early forties and slightly punkish. Theresa has a sleeve tattoo, while Brad has a ZZ Top-style beard. More importantly, both are workaholics who answer work-related text messages on a Sunday.

'Sure – what's the deadline?' she replied, while eating enchiladas.

'Monday evening or Tuesday. We need to write a digital plan for the campaign.'[2]

Every political campaign now has a 'digital plan'. It's industry talk that refers to the gurus, content producers, targeted ads and eye-wateringly large numbers that now feature in every election. We'll never replace door-to-door canvassing, which studies find is still the most effective technique to persuade voters, but no one serious runs elections without a digital plan these days. Brad's plan was to make the campaign the most data-driven in history: to take the philosophy of Silicon Valley and apply it to politics. Out with intuition and gut feeling; in with testing, measurement and scientific precision. He knew they would raise less money and have less support from the media or beltway pundits than the formidable Clinton machine, the likely opponent. So he decided he would use data to 'hack' the election.[3]

Once the nomination – and the contract – was secured, Brad's team set up shop in a nondescript San Antonio building, just off a busy freeway, intentionally out of the spotlight. He reported to Jared Kushner, who ran the campaign. 'It started up as four people in a room, and Brad saying "make cool stuff",' Theresa said later. It grew rapidly, and they soon took over the whole third floor of the building, adding cafeteria tables to the large, empty rooms.[4] The Republican Party heavyweights moved in, including Gary Coby, head of advertising for the Republican National Convention (RNC). So did Cambridge Analytica, the UK data analytics firm, who sent thirteen staffers led by chief product officer Matt 'Oz' Oczkowski, who had enormous biceps and a habit of walking around the office carrying a golf club. 'One of the smartest motherfuckers I ever met,' Theresa wrote about him later.[5] The department soon become known as 'Project Alamo'; as the campaign got into full swing, several dozen people, short of sleep and fuelled by pizza and Dr Pepper, relentlessly bombarded millions of Americans online with pro-Trump content. It was the largely unseen front line in the most peculiar election in living memory. More than an election, this was an information war.

Project Alamo

The data-led approach to elections pre-dates the internet – the Republican Party boasted in the 1890s that it possessed a complete mailing list of voters, with names, addresses and ages.[6] But, as we've moved online, the political campaigns have followed us there. For decades political parties have been building up increasingly detailed insights using shopping data, web browsing history and voter records to help with their targeting and messaging. In 2008, for example, analysts working for Barack Obama assigned a pair of scores to every voter in the country that predicted how likely they were to cast a ballot, and whether they supported his campaign.[7] Hillary Clinton, too, had an extremely sophisticated system of targeting voters online.[8] Every election now is a mini arms race.

And this time the Republican Party turned to a company, Cambridge Analytica, in order to get the edge on the opposition.

It was not a coincidental choice. One of Cambridge Analytica's key investors is the billionaire businessman and Trump backer Robert Mercer, a famously reclusive computer programmer who made his fortune as co-chief executive of the New York-based hedge fund, Renaissance Technologies. RenTech, as it is known, uses big data and sophisticated algorithms to predict trends in global markets and place winning bets on them. In this world tiny gains, a fraction of a per cent here or there, can yield huge rewards. In 2013 Cambridge Analytica was set up as an offshoot of a company called 'Strategic Communications Laboratories' (SCL), which had extensive experience in branding and influencing public opinion, specialising in military and intelligence psychological operations, or 'psy-ops' – tasks like persuading young men not to join Al-Qaeda. The idea was to figure out how to apply these techniques to politics – and especially to help the Republican Party, which Mercer felt had fallen behind the Democrats in their digital campaigning.[9] Mercer invested a load of money into the new company. Cambridge was also part of a tight pro-Trump network: Steve Bannon, until recently boss of *Breitbart* and Trump's first head of strategy, was also a board member of Cambridge Analytica until he joined the administration.

From their inception Cambridge Analytica followed the Mercer bible. They built up a database of around 5,000 data points on some 230 million Americans. Some of the data was purchased from commercially available sources – web browsing histories, purchasing, income and voter records, car ownership and so on – and some was collected through Facebook and telephone surveys.[10] They were initially part of Ted Cruz's campaign for the Republican nomination, but once he dropped out of the Republican race, the company transferred to Trump. They brought their data to Project Alamo, and the RNC threw their own massive dataset – known as Voter Vault – into the pot too, and got to work.

Cambridge's main role inside Project Alamo was to use this data to build what they called 'universes'.* Each was a key target group for the Trump campaign, such as American mums who hadn't voted before and were worried about

* It's not exactly clear what data went into these 'universes' – and how much of it was Cambridge's own data, or how much was RNC data. According to Parscale, much of Cambridge's input was their analysis of other data.

childcare; pro-gun males living in the Midwest; Hispanics who were worried about national security, and so on. Dozens of these highly focused universes were created – and their members were modelled in terms of how 'persuadable' they were. It might seem odd to build categories like this based on spending patterns or web browsing history but, as I showed in Chapter One, that's how big data analysis works. With enough data, you can build up a surprisingly detailed account of someone. For example, Cambridge Analytica discovered during the electoral race that a preference for cars made in the US was a strong indication of a potential Trump voter.[11] So if consumer data records showed someone had recently bought a Ford but the RNC data revealed they hadn't voted for years, they should be ranked as a highly persuadable target.

Everything about the Alamo was data-driven, and mostly built around these universes. Presidential elections in America use an electoral college system – each state has an allocated number of college votes based on population size, and the winner of the state takes them all. To become president, candidates need 270 electoral college votes. Project Alamo's analysts identified 13.5 million persuadable voters in sixteen battleground states, and modelled which combinations of those voters would yield a winning number.[12] From that, a computerised dashboard offered recommendations about rally locations, which doors to knock and where to direct emails, letters and TV advertising.

The largest room in Alamo was called 'the bull pen'. This is where Theresa and her 'creatives' worked. Much of Theresa's day was taken up by designing what people like her call 'content'. Matt Oczkowski would tell her what each universe cared about, and she would tailor something for them. 'The data drove the content and it was a great marriage,' Theresa later said. Alamo tested their messaging relentlessly. Gary Coby sent out multiple versions of fundraising emails and thousands of versions of Facebook ads and quickly worked out which performed best. They tested web pages for donations with red buttons, green buttons and yellow buttons. They even tested which unflattering picture of Hillary performed best.[13]

In 2017 I visited Project Alamo to interview Theresa for the BBC series, *The Secrets of Silicon Valley*. It was the middle of the summer in Texas, so indescribably hot. I flew into nearby Dallas and drove three hours to San Antonio to find Theresa waiting for me outside that tall, nondescript building off a busy freeway. I was the first journalist to be allowed inside the building, Theresa said, although by then it was completely empty. Theresa walked me from vacant room to vacant room, reminiscing about the all-nighters during the campaign. After my tour she pulled out her laptop and showed me some of the ads she'd designed and sent out into the world. One such ad was aimed at a universe defined by Cambridge Analytica as 'working mothers concerned about childcare'. It was the usual shtick – a soft voice narrating, the presentation of a happy-but-concerned family, and the message that Trump is worried, just like you. But Trump himself was absent. 'This is warm and fuzzy,' said Theresa. 'For that audience there, we wanted a softer approach.' For other universes, Trump was front and centre.

This relentless arms race using sophisticated big data techniques is not going to slow down. Every election is becoming datafied in this way – spread by a network of private contractors and data analysts who offer these techniques to political parties all over the world. Several months before Trump's victory, for example, the group campaigning for the UK to leave the European Union took a very similar approach. A few months after the referendum, Vote Leave's campaign director Dominic Cummings wrote a handful of long blogs explaining why they won. Although he rejects any single 'why', it's clear that he thinks data was instrumental:

> One of our central ideas was that the campaign had to do things in the field of data that have never been done before. This included a) integrating data from social media, online advertising, websites, apps, canvassing, direct mail, polls, online fund-raising, activist feedback and some new things we tried such as a new way to do polling … and b) having experts in physics and machine learning do proper data science in the way only they can – i.e. far beyond the normal skills applied in political campaigns. We were the first campaign in the UK to put almost all our money into digital communication then have it partly controlled by people whose normal work was subjects like quantum information … If you want to make big improvements in communication, my advice is – hire physicists, not communications people from normal companies.

Just like Brad, Cummings set up Vote Leave like a Silicon Valley start-up, with physicists, data, innovation and constant testing of ads or messages. One especially smart move involved inviting people to guess the results of all 51 matches in the Euro 2016 football tournament with the chance of winning £50 million, in exchange for their phone number, email, home address and a score of 1–5 in respect of how likely they were to vote for staying in the EU.[14] This, of course, fed into the models.

Cummings estimates that they served up around one billion targeted adverts during the Brexit campaign, mostly via Facebook (they spent £2.7m with a company called AggregateIQ, who specialise in targeted Facebook adverts). Like the Trump campaign they ran many different versions, testing them in an interactive feedback loop.*[15]

The evolution never stops. In the 2017 UK general election, the Labour Party took a different approach, although the overall aim – to change the information environment – was the same.[16] Rather than sponsored ads, Jeremy Corbyn's fans produced

* Carole Cadwalladr wrote several long features about Cambridge Analytica for the *Guardian*, about its involvement in the Vote Leave campaign – the company is currently suing the paper. To make matters still more complicated, at the time of writing the Information Commissioner's Office (ICO) is investigating the use of data analytics for political purposes.

huge volumes of 'organic' content themselves and shared it in tightly networked groups, meaning their messages – real things written by real people – reached far more people and were more believable than they would otherwise have been. There was also an ecosystem of left-wing 'alternative news' outlets that churned out widely shared and hyper-partisan pro-Corbyn stories. Corbyn Snapchatted during a brunch with the rapper Jme – it seems unlikely that this was an idea that he came up with himself. One of Labour's videos, 'Daddy, why do you hate me?' was a fictional conversation between a little girl and her dad set in 2030, about why he had voted for Theresa May. It was emotive, misleading, mawkish and potentially offensive – and viewed millions of times in two days.

Labour also relied on the technical stuff, and quietly but effectively used data modelling to identify potential Labour voters, and then test them with messages.[17] They used an in-house tool called 'Promote' which combined Facebook information with Labour voter data, allowing senior activists to send locally based messages to the right (i.e. persuadable) people.[18]

The key to understanding why these tactics can be so effective was revealed a few years back, almost by accident. During the 2012 US presidential election, millions of voters told the world about their little civic act by posting *I voted* on Facebook. The company worked out that friends who saw these posts were themselves slightly more likely to vote as a result – so much so, in fact, that Facebook may have increased turnout by 340,000 people. Bear in mind that the presidential race in 2000 was won by just 537 votes – if Facebook had showed 'I voted' posts to potential Democrat voters in Florida in that election, it might have swung the entire election. According to Robert Epstein, a psychologist at the American Institute for Behavioral Research and Technology, based on the win margins in national elections around the world, Google could determine the outcome of 'upwards of 25 per cent' of them based on how it displayed search results.[19] There is no evidence that Facebook or Google have or would do such a thing, intentionally or otherwise – but it does illustrate that whoever controls information has immense power, and that even small changes in the online environment can be critical.

Facebook, in case you've not been paying attention, is a highly effective mechanism for advertisement delivery, because of how finely grained it can target users (one technique in particular, known as Lookalike Audiences, is highly regarded among those in the know).[20] Both Corbyn and Vote Leave relied heavily on Facebook as a mechanism to reach audiences.[21] But neither used it as much as Brad Parscale did on the Trump campaign. Over the course of the campaign Alamo spent around $70 million on Facebook advertising, running up to one hundred adverts a day, and often thousands of versions of each, constantly tweaking to see which version performed best.[22] Brad told *CBS* in October 2017 that Facebook made the difference, allowing him to reach people who had previously been unreachable. 'It lets you get to places you'd never get to with TV ads.'

I have run Facebook adverts myself. Back in 2010 I used Facebook to target ads at supporters of radical right-wing political parties in Europe, asking them to fill in a survey for my research organisation, Demos. It's not easy. Big-spending clients therefore sometimes get help from Facebook directly. Brad told CBS *60 Minutes* that he emailed Facebook and Google, asking for embedded staff – and even insisted that they were Republicans. 'I want to know every single secret button and click you have', he told them. 'I want to know everything you'd tell Hillary's team, and then some.' Sitting in Alamo, alongside Cambridge Analytica, were seconded staff from Facebook and Google, whose job it was to ensure Trump got the most bang for his buck. I know this because Theresa pointed out where they were sitting, and couldn't sing their praises highly enough. '[Facebook] gave us the white glove treatment', she told me as we walked around. 'They were our hands-on partners, as far as being able to utilise the platform effectively.'[23]

I was surprised when Theresa told me that social media employees – and ones who shared the campaign's political views – were working directly with the Trump team, but perhaps I shouldn't have been. By now we've all got used to the idea that sophisticated cookies and tracking software follow us around the web. But this isn't only to bombard us with holidays, make-up or jeans: it can be used just as easily to promote politicians. We are put unwillingly and unknowingly into 'buckets' or 'universes' by clever data analysts who obsess over 'click through rates' and 'conversion'. For campaign managers we are 'targets' to be 'hit' with political content. We used to call this sort of thing propaganda. Now we call it 'a behavioral approach to persuasive communication with quantifiable results', and give awards to the people who are best at it.[24]

Left unchecked, the continued evolution of these techniques will change how we form political choices, what sort of people we elect, and even whether we think our elections are truly free and fair.

Modern mass-party politics has always been about programmatic offers – broad-based appeals that could build large alliances. This is important, because as the social scientist Francis Fukuyama argues in *Political Order and Political Decay*, political parties with broad programmes allow citizens with different and varied interests to collectively organise and shape policy. The alternative is squabbling, divisive special-interest groups. (This also helps citizens who are on the losing side to accept defeat, because they know they might win next time.)[25]

Big data, however, points to a more personalised model: work out who people are, find the one thing they care about, and zero in on that. Persuasive adverts have always been used in politics – remember 'Labour isn't Working'? – but instead of sending out a mass advert to millions, campaigns can now target a specific set of voters, each with specific promises and pledges, based on what they already care about.

This is a radical change with far-reaching consequences. It is important that everyone receives the same message – or at least knows what others are receiving. That's how we are able to thrash out the issues of the day. If everyone receives personalised messages, there is no common public debate – just millions of private ones. In addition to narrowing the scope of political debate (research suggests that candidates are more likely to campaign on polarising issues when the forum is not public), this will diminish political accountability. Hyper-personalisation incentivises politicians to make different pledges to different 'universes' of users. But how can we hold anyone to account if there is no clear, single set of promises that everyone can see and understand? And how do we even know if we're getting the real Trump anyway? When I was at Alamo, Theresa told me that *she* wrote many of Donald Trump's Facebook posts. That was odd. I'd always assumed Trump wrote his own posts. I'd read many of them, and they certainly sounded like him. Nope, it was Theresa, sitting in her San Antonio office. 'I channelled Mr Trump,' she told me, smiling. 'How do you channel someone like Donald Trump?' I asked. 'A lot of believe mes, a lot of alsos, a lot of verys ... he was really wonderful to write for. It was so refreshing. It was so authentic.' She seemed unaware of the irony.

Personalisation causes problems for regulators too, of course. Because ads are so personalised, and delivered to unique users, it is more difficult to check whether they are accurate. UK law prevents candidates from making false claims about each other. But Facebook allows people to use so-called 'dark posts' – non-public posts that only the targeted people can see, and quality assurance is extremely hard.[26]

In the mad dash to get an edge, each political party rushes to adopt the latest techniques, rarely considering where it might take us all. Several journalists – myself included – have become mildly obsessed over whether Project Alamo used one specific micro-targeting technique known as 'psychographics'. This is the stuff Kosinski showed me in Chapter One: trying to figure out people's personality traits and designing adverts based on that. Cambridge Analytica have used this technique in the past, and claim that they can predict the personality type of every single adult in the US. They tried this during their work on the Ted Cruz campaign, though it's not clear how well any of it worked.[27] Then, in March 2018, a former Cambridge Analytica whistleblower told the *Observer* that a major part of company's powerful data sets were derived from Facebook data they had accessed in an improper fashion. The resulting furore led to several days of front page media coverage, the UK's Information Commissioner seeking a warrant to look at Cambridge Analytica's databases, and billions of dollars being wiped off Facebook's value.[28]

Shortly after returning from San Antonio I managed to secure an interview with Cambridge

Analytica CEO Alexander Nix. As I walked in to the ordinary looking office in central London – all offices are normal looking, except those of tech firms – I spotted a framed posted with a picture of Trump and a quote from famed US pollster Frank Luntz: 'There are no longer any experts except Cambridge Analytica. They were Trump's digital team who figured out how to win.' Rows of employees were sitting staring at screens: project managers, IT specialists and data scientists.[29] On a shelf in Nix's glass office were copies of *The Bad Boys of Brexit*, the book written by UKIP donor Arron Banks, and *Stealing Elections* by John Fund. He seemed perfectly happy with these techniques, and said that micro-targeting was just getting started and represented the future of campaigning. 'It's going to be a paradigm shift ... and that is the way the world is moving.' I asked whether Nix used psychographics during the Trump campaign, and he denied it. So did Brad Parscale, in his *60 Minutes* interview.[30] (Cambridge Analytica also strongly deny allegations that they obtained Facebook data illegally or used it without the proper permissions.)

I understand why people get nervous about psychographics, because the idea feels extremely manipulative. And of course it matters that data is harvested and used legally and ethically. But in one sense this is a distraction. The bigger picture is the way that companies like Cambridge Analytica understand our inner thoughts, rather than a distinct technique.* After all, just imagine what personality targeting will be possible with 'the internet of things'. There are lots of stories these days about how internet-enabled devices present a security risk – like your fridge or baby monitor getting hacked. But think about what the explosion of everyday life data will do for political campaigns. Consider it: within a decade your fridge data will know what time you eat, your car will know where you've been, and your home assistant device will work out your approximate anger

* 'Did you use psychographics in the Trump campaign?' I asked Nix during our interview. 'No, we didn't have time,' he said. 'Did we build specific psychographic models for the Trump campaign? No we didn't,' he added. I pressed further, and in the end he conceded that the Trump team had taken 'legacy' data with them from Cruz's campaign to Trump's. 'We took models we'd made previously, and integrated them.' I'm still not exactly sure what this means. The best I can conclude is that their data points – some five thousand on 250 million people – had become so complicated that Cambridge Analytica couldn't easily untangle its own data. They may not have explicitly tailored each advert, but within their data there were some personality traits. I thought this was quite a nice scoop at the time. The journalists that followed this all said it was a step forward.

levels by the tone of your voice. I guarantee this data will be gobbled up by political analysts. By cross-referencing fridge data against the number of emotional words in your Facebook posts, Cambridge Analytica or some other company will correlate that you're more angry when you're hungry. Further analysis will calculate that people who are angry are more likely to vote for 'law and order candidates'. Armed with your fridge data, smart car data, work calendar data and Facebook data, your smart TV will fire a personalised, crime-related ad at you just at the moment you're starting to feel peckish.[31]

I've no idea where it will end. Give it a few years and, just as you're relaxing in some virtual reality paradise, a Trump avatar-bot will roll up, and know precisely how to press your buttons.

In the long run, the constant A/B testing and targeting like this might even encourage a different *type* of politician, because it promises to turn politics into a behavioural science that relies on triggers and nudges rather than publicly aired argument.[*32] It is reasonable to assume this approach would most

---

* Interestingly, some research shows that candidates are more likely to campaign on divisive issues when the forum is not public – which suggest these techniques could drive greater polarisation.

help politicians with flexible campaign promises, the ones who flap around in the breeze, make hundreds of contradictory statements, and change their minds at every propitious moment, because that creates more content for people like Theresa to package up and sell to voters.[33] Perhaps the politicians of the future will be those with the fewest ideas and the greatest talent for being non-committal and vague. I can imagine a campaign team asking their candidate to pre-record hundreds of contradictory messages, which they could then fire at different audiences. If every voter is a data point who receives, not messages from politicians, but a perfectly targeted machine-generated advert, finely tuned and retuned to suit a particular personality and mood, an algorithm which runs itself and improves iteratively, without making any serious effort to engage with you – then elections will become little more than software wars.

But the more politics becomes a question of smart analysis and nudges rather than argument, the further power will shift away from those with good ideas and towards those with good data and lots of money.

It turned out that Project Alamo was a small piece of a much bigger puzzle in which influential people battled over the shape of reality. Robert Mercer had also invested in *Breitbart News* – best described as a right-wing *Huffington Post* that specialises in stories castigating liberals, bad Muslims, and the 'mainstream media' – which became a highly influential source of anti-Clinton and pro-Trump news.

According to the academic Jonathan Albright, the US election was dominated by a 'micro-propaganda machine', a network of thousands of web pages from the radical right hyper-linking to each other and spreading 'false, hyper-biased, and politically loaded information'. Many used advanced tracking cookies that followed users around the web, advanced programmatic ad delivery and AI content optimisation to serve up more conspiracy theories to the so-inclined.[34]

It is increasingly clear that Russian president Vladimir Putin was engaged in this information war too. For some years, the Russian Government has known that covert media manipulation online can subtly shift public opinion in ways that promote its interests – supporting far-left and far-right parties across Europe and firing up campaigns of internet disinformation throughout the Ukrainian crisis. During the US election the Russian Government took these Cold War techniques up several notches. Thousands of paid content producers pushed out pro-Trump or anti-Hillary content, flooding feeds and overwhelming serious hashtags with nonsense, making them unusable. Russian hackers ran very big Facebook pages, which created the illusion of grass-roots support for Trump. They allegedly hacked Hillary Clinton's private emails and shared them with the whistleblowing site WikiLeaks – who leaked them slowly over the campaign, and to good effect. They also ran an aggressive campaign of paid advertising on Facebook and Google.

I won't tell this story in full here, because it is still unfolding (at the time of writing, the investigation into alleged collusion between the Trump campaign and the Russian Government is ongoing).* But it seems that the purpose was obviously the same as Alamo: to win the information war, shape people's reality and use the internet to subtly shift opinion in new and hidden ways.

---

* Twitter revealed the handles of some 36,746 Russian-linked 'bots' (automated accounts) which tweeted a total of 1.4 million times in the two months before the election, and were viewed almost 300 million times. (Other researchers have placed the number far higher.) According to researchers at the Oxford Internet Institute, pro-Trump bots outnumbered Clinton's five to one – and they were carefully timed to flood pro-Clinton hashtags during the debate and then stopped right after election day.

Importantly, the Russian meddling didn't always display a pro-Trump agenda. Just as often, the aim was to sow discord and disharmony more generally.[35] After the shooting at Marjory Stoneman Douglas High School in Parkland, Florida in February 2018, Russian bots and trolls started posting inflammatory content about gun control *on both sides of the argument.* The same thing happened after shootings in Las Vegas, the NFL protests and high-profile news stories about sexual misconduct. According to former CIA Director Mike Pompeo, this now constitutes a 'serious threat' to democracy – not because it might decisively swing an election, but rather because it chips away at social cohesion and public confidence in the democratic system itself. The Kremlin doesn't care what the US law is on gun control – but if the American people are arguing, the Russian government believes it is winning.

The scale of the Russian disinformation effort was staggering, but hardly surprising. Democracies with free media, fair elections and an open internet are more subject to international meddling than closed autocracies (and if some of the projections I set out in the next chapter about future unemployment are correct, 'paid content producer' whose job it is to influence online opinion might one day be a very desirable position). To their credit, the tech firms – especially Facebook – rushed to promise action after this was revealed, restricting political ad purchasing and hiring more people to manually review the content. Twitter created an 'Advertising Transparency Center' to show how much money each campaign spent on advertising, the identity of the organisation funding the campaign and what targeting demographics were used. Mark Zuckerberg seems to have had a Damascene moment towards the end of 2017, when he acknowledged that the company needed to behave more like a responsible publisher that takes editorial decisions, rather than as a neutral platform that treats all information equally. This will certainly help. There are also measures governments can take too, such as bringing election laws up to speed, which I discuss at the end of this book.

But even this will not eradicate the problem entirely, because a networked world where everyone is posting from everywhere all the time is simply impossible to control completely. This is more than Russian influence: democracies can no longer effectively police their information borders. Facebook's dream to connect the world also means connecting Russian bots with British voters and gullible news outlets, fake news purveyors with floating voters, and Theresa Hong with worried American mothers who have never voted before.

Every election has become an arms race, and the problem with arms races is that they are very difficult to slow down. Big Tech has built the infrastructure for selling stuff – some of the most sophisticated and connected configurations man has ever dreamt up – and now these infrastructures have been repurposed to win elections. In the red corner: a multi-billion-dollar business of influence and control which gets more accurate and targeted every year. In the blue corner: a handful of weak and dated election rules designed for the era of mass broadcast and door-to-door canvassing.

The night of the 2016 election started well for Democrats – early exit polls looked good, and analysts confidently predicted a comfortable win for Hillary Clinton. David Remnick, editor of the *New Yorker*, drafted an essay about the country's first woman president. Producers at Fox News predicted they'd be calling the race for the Democrats before midnight East Coast Time. Even the Republican Party seemed to be preparing for a round of buck-passing.[36]

But as the evening rolled on, there were signs that things weren't going according to the script.

Votes in Florida were taking longer than expected to count, and in a handful of early reporting precincts there were more pro-Trump votes than pollsters had anticipated. Turnout in Ohio among white working-class voters was rumoured to be high. Michigan and Wisconsin still hadn't called. Surveying all this, David Chalian, Political Director at CNN, told his producer Terence Burlij at 9.15p.m. that he thought Trump might actually win. 'He looked at me like I was crazy. You could sense that the night was different.'[37]

After weeks working in San Antonio, Brad Parscale had decamped to Trump Tower in New York for the results and was closely reviewing every scrap of news. Darrell Scott, a member of Trump's transition team, found Brad on the fourteenth floor, running through scenarios on his laptop. 'How we looking?' he asked. Brad told him they'd over-performed almost everywhere, pointing at the screen.[38] Darrell texted Matt Sheldon, a Republican publicist. 'The computer guys are already saying that he's going to win,' he wrote. 'Parscale's throwing a paper airplane right now across the room.'

As the mood was lifting in Trump Tower, the atmosphere at Clinton HQ was very different. Aides stopped doing live on-location interviews and started frantically calling contacts in key states, to figure out what was going on. At 10p.m., the TV monitors in the Clinton press room switched from running the cable news feeds back to old promo material. 'That felt like a turning point,' said one CNN producer who was present.[39]

At approaching 11p.m., the result everyone was waiting for was finally announced: Trump had won Florida, a key battleground state where he had no ground game and had been polling badly. Soon after, Ohio, another swing state, went the same way. Analysts, not for the first time that night, recalculated Clinton's road to victory. She needed to win Pennsylvania (worth 20 electoral college votes), Michigan (16) and Wisconsin (10) in order to reach the magic number of 270. 'Her path is getting narrower and narrower,' CNN anchor Jake Tapper told viewers.[40]

According to Jim Margolis, who'd been a senior advisor on both Obama campaigns, people inside the Clinton war room were phoning their people on the ground in Wisconsin and Michigan to figure out why those states hadn't yet declared for Clinton when all the polls had pointed to an easy win for her.[41] After all, these states had gone Democrat in the last six presidential elections. If they held, she was still in the game.

Except that a few months earlier Brad, sitting inside Project Alamo, reviewing Cambridge Analytica's universes, had realised they were winnable. The models suggested there were enough swing voters and non-voters who could be persuaded to vote for Trump. He shifted budgets around to focus on these Rust Belt states.[42] 'I took every nickel and dime I had out of everywhere else, and I moved it to Michigan and Wisconsin,' he later told *60 Minutes* on CBS. Jared Kushner told Trump to start campaigning in Pennsylvania, too. At the time several pundits said this was mad – it was on the wrong side of the much-vaunted 'blue wall' of solid Clinton territory. But Brad was following the data.

At approaching 2a.m. East Coast Time, Trump won Pennsylvania, pushing him to 263 electoral college votes, and the door for Hillary was closing fast. Half an hour later AP News projected that Trump had won Wisconsin – taking him across the victory line. It seemed apt that it was Wisconsin, the state that no one thought possible except the data guys inside Alamo. Trump was the first Republican to carry the state since Reagan in 1984. A few minutes later Clinton called Trump and conceded. She'd won the popular vote by two million votes, but she didn't win in the places where it counted.

As Trump took to the stage for his victory speech a couple of hours later, Brad – who at six foot five towered over the delighted crowd – looked out over the assembled supporters. He glanced over at Darrell Scott and simply said, 'I told you'.

Far too many otherwise-intelligent people, unable to comprehend Trump's popularity, believe that voters were duped by Brad or Theresa, or even by Vladimir Putin, into ticking the box for Trump. Those involved are happy to propagate this myth, because it's good for business. Ever since Cambridge Analytica were credited, in multiple outlets, as the geniuses behind his victory, trade has been booming. 'It's like drinking from a fire hose,' Oczkowski said in a recent interview. 'Aside from Antarctica, we've gotten interest from every continent.'[43]

The truth is less straightforward. Obviously many factors led to Trump's win – economic stagnation, his dreary opponent and the white working-class revolt. And as Richard Hofstadter famously wrote, there is a 'paranoid' style in American politics, which stems from the fear that shadowy, powerful interests are doing the Republic down.[44] There's certainly some partisanship at play, too. I don't recall similar levels of outrage when it was revealed in 2012 that President Obama's team had placed voters into 30 buckets and ranked them according to persuadability, and that Google's Eric Schmidt advised the campaign. Liberals were apparently extremely comfortable with the idea when it was their guy doing it. That was a mistake.

But, in a relatively close race with two unpopular candidates and a small number of key marginal districts, Project Alamo probably *was* decisive. Brad's decision to bet the house on digital, Cambridge Analytica's refined universes and the hands-on help from Facebook all meant Trump could reach enough of the right people in the right districts with the right messages at the right time. Throw in a load of trolls and bots nudging the online debates his way, and that was enough to swing it. When the final counts were made public, it was revealed that Trump won Pennsylvania by 44,000 votes out of six million cast, Wisconsin by 22,000 and Michigan by 11,000. These are tiny numbers – less than one per cent of the votes. If they had gone to Clinton, as projected, she would have been elected president.

Not all elections will be this close. But, soon enough, nearly all will be run with similar combinations of big data, algorithms, granular targeting and supposedly organic and authentic content. This isn't a story about Trump 'stealing' an election.

Who wins and loses is less important than whether the integrity of elections themselves are at risk. Elections are comprised of hardware and software. The hardware is the technical rules by which people get to have their say on who governs them – accurate counts, polling stations, a means to register as a candidate and so on. But elections also depend on software: people should be left to make their minds up freely and with a clear head, based on a sound understanding of their interests and accurate information. If some people can unduly influence that election software in ways that we are barely aware of, then elections aren't really free and fair. Unless we can understand the techniques employed and hold those who employ them accountable, there is a chilling prospect that whoever owns the data also owns the future, because they can hack the software – and this might just be enough to make a difference. Meet the new boss. Same as the old boss. But now armed with algorithms and big data.

Chapter 4: Driverless Democracy

What Happens to Citizens When AI Takes All the Work?

Science fiction is fast becoming science fact, as rapidly improving artificial intelligence starts to impact our economy. However, rather than speculation about a 'jobless future', we should be worrying about growing inequality and whether the coming tech revolution will wipe out the middle class.

LIKE MANY SILICON VALLEY start-ups, Starsky Robotics was founded by two twenty-somethings who regard sleep as optional. Any successful new tech firm needs someone who understands technology and someone who understands business, and these two distinct skills are rarely found in one individual. At Starsky, Kartik Tiwari is the 'tech guy' who specialises in artificial intelligence, and Stefan Seltz-Axmacher is the 'serial entrepreneur' guy, who mostly specialises in starting-up start-ups. Neither, you'll note, specialise in trucking, which was the domain of their company. This did not seem to bother investors, since this 11-person business, with ambitions to revolutionise the entire trucking industry by building a self-driving fleet, managed to raise millions of dollars of funding from venture capitalists.

'Everyone thought I was mad,' 27-year-old Stefan told me when I visited Starsky's Florida headquarters, a large rented property in a gated community, a few months ago. These days however, like many other industries, trucking is being disrupted by data science, artificial intelligence and venture capital.

Stefan had agreed to let me drive in Starsky's newest and shiniest truck with their resident driver Tony Hughes, a diminutive and friendly man with 20 years' experience, who is perhaps better described as part-driver, part-machine supervisor. Tony is in his fifties, with a high school diploma in general studies from Shawnee Mission Northwest (Kansas) and a 'solid track record of achieving efficient, cost-effective transportation operations', but now finds himself training the machines that might eventually put him out of a job. He's spent months driving the Starsky truck up and down certain routes, over and over, so the software can collect data on how he does it. From this experience, it 'learns' how he behaves and how to mimic him. The law requires a human in the cab, and anyway the software is still in 'dev mode', so Tony supervises – there needs to be someone responsible in the cab, in case something goes wrong. Neither Stefan nor Kartik, both wannabe truck magnates, have any idea about how to actually drive one.

Figuring they might as well make some money rather than just transport an empty truck in circles around the country, Starsky vehicles sometimes make actual deliveries while they're training the software. So I jumped aboard 'Rosebud' (everyone names their trucks, apparently) in Orlando, Florida along with Tony, Stefan, Kartik and 5,000 pounds of empty milk cartons. Our destination was a depot in Deerfield Beach, around 200 miles south. Underneath Tony's feet and behind his oversized wheel were work-in-progress wires, pumps, shiny levers and many cogs. They were connected to computers in the back of the cab, which were under Kartik's command. The software controlled the pedals and steering wheel, which constantly adjust to real-time data collected by mounted radars and computer vision sensors that covered the vehicle: position, speed, road markings, other cars' positions and speed and so on.

We left the narrow residential roads, and joined Freeway 95.

Tony turned to Stefan: 'I could kick the system on if you guys are ready.'

'Rosebud on,' Stefan shouted into his walkie-talkie to other crew members, who were following in a car.

Tony flicked a little blue switch, and we went from 'manual' to 'auto'. The truck made a tiny lurch, as if to unshackle itself from Tony, who stretched back and crossed his legs, relaxed. 'I trust Rosebud completely,' he told me. 'I've trained her for months. She can almost drive as well as me now.'

It was, of course, terrifying and exhilarating to thunder along in a 40-tonne HGV without a driver. As the first major bend in the interstate approached, I did what I do when I'm on a flight with turbulence, and looked at the crew. Tony seemed calm. Rosebud took the corner smoothly, of course. I slowly relaxed, and learned that nothing really happens on the long, straight, wide American freeways anyway. 'They're ideal roads for machines,' Stefan told me as we chugged along. Around 70 per cent of our journey to Deerfield Beach was operated entirely by machine. I soon became a little bored, as novelty gave way to tedium. This felt appropriate; some engineers worry about the risk of 'vigilance decrement', as humans lose practice and become less able to deal with emergencies. My tedium in turn gave way to the dawning realisation that what I'd thought was sci-fi was in fact fast becoming sci-fact.*

In 2004, respected AI-researchers from the Massachusetts Institute of Technology (MIT) concluded that autonomous vehicles were a pipe-dream, driving being a skill that required too much human intuition and motor skills.[1] But we should never underestimate the speed at which digital technology can advance. Millions of dollars of investment are now pouring in from Uber, Google, Tesla, Mercedes, Volvo, Starsky and others. Several countries (including the UK) are encouraging 'real world' testing, and the British Chancellor expects that driverless cars will be on the road by 2021. The way it's going, regulations

* The software in most driverless vehicles works in roughly the same way: sensors can 'see' what a driver would, and are programmed to respond: principally to road markings, other cars and pedestrians. Unlike humans, they have no blind spots. In fact, most software is also able to work out what it thinks other drivers' blind spots are, and keep away from them. Most of the 1.3 million people who are killed each year in car accidents die due to driver error. We will not remember this when there are fatalities in driverless cars, and there will inevitably be a backlash about how dangerous they are.

and insurance issues are more likely to slow this down than the technology.

A world without work?

Driverless vehicles are just one application of an artificial intelligence revolution that is sweeping through the global economy. The leaps forward in AI – which were first gradual and then sudden – are behind the first genuine mass panic of the twenty-first century: that we are entering a world in which robots will take every job currently done by humans, putting us all out of work. Media outlets seem to enjoy writing scary headlines about this.*

There are a couple of widely held misconceptions about AI that need to be cleared up. Despite the Hollywood movies and the breathless headlines, no machines are remotely close to reaching a human level of intelligence, which we can define as 'performing as well as humans in a series of different domains' (often known as 'general AI'). Although divided, most experts don't think this level of intelligence will be possible for another 50 to 100 years

* 'You Will Lose Your Job to a Robot' predicted *Mother Jones* while the *Guardian* confidently informed us that 'Robots will destroy our jobs'.

– but to be honest, no one really has a clue. And whether machines will ever achieve consciousness is altogether another question entirely, and probably one best left to philosophers rather than roboticists.

The obsession with marching machines, *Terminator 2*'s Skynet and humanoids unhelpfully distracts from the real action, which is 'domain-specific' AI and often uses a technique known as 'machine learning'. A human feeds an algorithm with data and teaches it what each input means. From that it can detect patterns, from which it can mimic a particular human behaviour or perform a very specific task, whether driving up the interstate, predicting the weather, giving credit scores, reading license plates and so on.

Machine learning has been around for many years and is already embedded in many aspects of our economies, whether suggesting a next Amazon purchase or a new Facebook friend. ML relies on data to learn and, because we now produce so much of it, it has been able to grow quickly. The falling prices of computing have helped bring about a powerful self-reinforcing feedback loop. More data fed in makes ML better, which allows it to make more sense of new data, which makes it better still, and so on. More sophisticated ML is being developed all the time. The latest involves teaching machines to solve problems for themselves rather than just feeding them examples, by setting out rules and leaving them to get on with it. This is sometimes called 'deep learning', which attempts to mimic the activity that occurs in layers of brain neurons through a neural network to spot patterns or images very effectively.[2]

To understand how this is different and potentially more powerful than classic ML, consider the ancient Chinese game Go. Machines have been beating humans at chess for years, but Go is more difficult for machines because of the sheer number of possible moves: in the course of a match, there are more possible combinations than there are atoms in the universe. A few years ago, DeepMind, a Google-owned AI firm, built software to play the game, called AlphaGo. It was trained the 'classic' ML way, using thousands of human games; for example, being taught that in position x humans played move y; and in position a, humans played move b, and so on. From that starting point AlphaGo played itself billions of times to improve its knowledge of the game. In 2016, to the surprise of many experts, AlphaGo decisively beat the world's best Go player, Lee Sedol. This stunning result was quickly surpassed when, in late 2017,

Deep Mind released AlphaGo Zero, a software that was given no human examples at all and was taught the rules of how to win by using a deep learning technique with no prior examples. It started off dreadfully bad but improved slightly with each game, and within 40 days of constant self-play it had become so strong that it thrashed the original AlphaGo 100–0. Go is now firmly in the category of 'games that humans will never win against machines again'.

Most people in Silicon Valley agree that machine learning is the next big thing, although some are more optimistic than others. Tesla and SpaceX boss Elon Musk recently said that AI is like 'summoning the demon', while others have compared its significance to the 'scientific method, on steroids', the invention of penicillin and even electricity. Andrew Ng, former chief scientist at Baidu, reckons that there isn't a single industry that won't shortly be 'transformed'.

AIs are starting to outperform humans in a small-but-quietly-growing number of narrow tasks. Over the last year alone inroads have been made into things such as driving, bricklaying, fruit-picking, burger-flipping, banking, trading and automated stock-taking. Legal software firms are developing statistical prediction algorithms that can analyse past cases and recommend trial strategies. Tools to analyse CVs are now routinely used by companies to help them filter out obviously unsuitable candidates (for example, those who do not have the correct A Levels). Using complex data models, software can now make predictions on investment strategies. According to the consulting firm McKinsey & Company, which examined 2,000 work activities in 800 occupations, 45 per cent of tasks people are paid to do at the moment could be done by *currently proven technologies.* Similarly, the Bank of England recently suggested that up to 15 million British jobs might be unnecessary within a generation.

I don't take these predictions all that seriously. Many of these applications are still young. Every new technological revolution unleashes similar speculation, and it is often wide of the mark. Even our wisest heads get it wrong – back in the 1930s John Maynard Keynes believed that the UK was witnessing 'technological unemployment', as the ability of machines to take over jobs outpaced the economy's ability to generate new ones. We've had tech-led disruption before, and we have usually found new (and often better) jobs. After all, machines tend to drive up productivity, which in turn stimulates more investment and demand.[3] A recent analysis of the American workforce between 1982 and 2012 found that employment *grew* in several areas where computers were used (gaming, graphic design and programming).[4] And in many instances, productivity gains driven by technology won't mean fewer jobs, but rather improvements in current ones. When AI techniques transform medical diagnosis – within the next few years – it won't mean fewer doctors, but better patient care because our busy doctors won't need to spend hours staring at scans. We are terrible at predicting what the jobs and industries of the future might be. Millions of people are already employed in roles that didn't exist 20 years ago: web developers, app designers, Uber drivers, lifestyle coaches and a thousand other things besides.

Gunning down Interstate 95 thinking about these issues while Rosebud took care of the driving, it struck me that the real challenge might not be jobs, but inequality. Driverless trucks will create plenty of great new employment opportunities.* Starsky's employees are all highly educated, enthusiastic young people doing very good jobs: robotics specialists, engineers, machine learning experts.

* The UK Government reckons the industry will be worth £28 billion by 2035, although this strikes me as a ludicrous speculation, which will have been forgotten by the time it's proven to be completely wrong.

They have created these jobs, and good for them. If they succeed, however, they will also remove the need for other jobs. Not all drivers will vanish, of course (Stefan says some will sit in an office and control several trucks at once, taking over from the truck while they manoeuvre in and out of busy depots or around complicated roundabouts). Plus, in the US at least, there is still a shortage of truckers. But over the coming years, it's likely that we won't need anywhere near the number of drivers that we currently do.[5]

For Tony, trucking has been his profession since the mid-1990s. It is a difficult and lonely job. But for many Americans without a college degree, it is one of the best-paid jobs available – especially in the poorer states such as Iowa or North Dakota.[6] Three per cent of Americans are employed in a driving job of one sort or another.[7] Will the drivers who lose out take any of the snazzy new jobs created by Stefan and Kartik? A handful might; some, like Tony, are already training the software. Perhaps others will re-skill, and claw their way up to the winners' table. I was told repeatedly in the magical Silicon Valley bubble where everything is possible that unemployed truckers in their fifties could retrain as web-developers and machine learning specialists – a convenient self-delusion that no one really believes. It is far more likely that many truck drivers, without the necessary skills, will drift off to more precarious, piecemeal, low-paid work – perhaps becoming taxi drivers (assuming they still exist) or Amazon warehouse operators or Mechanical Turk labourers who are paid an hourly rate to train software or fill in surveys. Perhaps they could clean the machines that clean the machines that repair the driverless trucks that they once occupied.*

Routine and non

We should be reasonably confident that AI will result in forward leaps in productivity and overall wealth. The big question is how the spoils of that wealth will be shared out. Just because machines can beat other machines that beat us at board games, it does not follow that they can undertake every useful task that humans currently perform. Many of the 'thinking' things humans do, such as extremely complicated calculations, can be easily replicated by machines. They can undertake routine or predictable tasks at a

* An old joke goes: the factory of the future will have only two employees, a man and a dog. The man will be there to feed the dog. The dog will be there to keep the man from touching the equipment.

speed and rate of accuracy far beyond anything we can manage. Conversely, a lot of things we do unthinkingly, like picking up a deck of dropped cards or tying shoelaces, are far harder for machines. They are significantly worse than humans at dealing with unpredictable situations, especially ones requiring sensorimotor skills. This is sometimes known as Moravec's Paradox, after the roboticist Hans Moravec who realised that high-level reasoning often requires little computational power, but low-level sensorimotor skills need a lot.* Not to mention, of course, that humans still win hands-down on breadth and range of intelligence.

This has some important ramifications. It means that the jobs most at risk will be those involving the routine tasks that can be most easily be done by machines. The safest, and most likely to be created, work will be 'non-routine'. The strange thing about our economy is that non-routine jobs tend to be either very well or very badly paid. A specialist in machine learning at Starsky Robotics or Google performs a non-routine job, since it involves a lot of intuition, creativity and independent thinking in

---

* For now at least – plenty of robotics companies are working to overcome Moravec's Paradox, especially as computing power increases.

unpredictable situations. So does a gardener, carer or Deliveroo cyclist. It's the jobs in the middle – what you might call 'routine cognitive' jobs – that will be most at risk. If you are a train operator, a mortgage adviser, a stock analyst, a paralegal, a credit analyst, a loan officer, a bookkeeper, a tax accountant or a radiologist, you might consider retraining.

If we lose these jobs, we will be heading towards what David Autor, a labour market economist at MIT, calls a 'barbell-shaped economy', a kind of extreme inequality. Anyone with the skills, diligence, money or fortune to work closely with technology – and especially AI – will likely see huge leaps in productivity and wages. There will be plenty more low-paid, insecure service jobs that won't disappear due to automation, too. But with millions of people competing to wait on, care for and serve food to the winners of the great tech revolution, wages for those jobs will be driven down further. The labour market in 2030: either a well-paid job at Facebook, or the opportunity to deliver food on a bicycle on minimum wage to these busy and important people. But good luck trying to get a stable non-cognitive job as a local news journalist, paralegal, truck driver or tax accountant.

I don't wish to put AI in the dock for crimes as yet uncommitted. But this kind of tech-fuelled inequality is a familiar pattern. 'There is no economic law that says that all workers, or even a majority of workers, will benefit from technological progress,' write McAfee and Brynjolfsson in their influential book *The Second Machine Age*, which argues persuasively that, while other factors are of course at play – including globalisation – technological advance over the last 30 years has been the main factor behind growing economic inequality. Skilled workers, they explain, tend to benefit most from new technologies, while others fall further behind. In the US, productivity has been rising, shiny new buildings are being built and corporate profits are increasing, but average salaries are falling. In a similar fashion, productivity in the UK increased by 80 per cent between 1973 and 2011 (although it is still low by the standards of the Organisation for Economic Co-operation and Development) but the hourly compensation of the median worker went up by only 10 per cent in real terms. All over the world – including in socialist Sweden and Mittelstand Germany – top earners and top jobs have been doing just fine, while for a lot of people in the middle and bottom, earnings and wealth haven't increased at all in real terms since the 1970s.

There are other forms of inequality that no one is thinking about at play here too. As a general rule, technology empowers those who have either the money or the skills to take advantage of it. The more powerful the tech, the more powerful the tendency. So let's say that people start owning personal AI assistants, as I outlined in Chapter One. These AI assistants would help their owners identify the best prices, the cheapest holidays, the most sound legal advice, write the strongest CVs and so on. Those who can afford the best AI assistants will see their prospects rise further, while those without will lag further behind. What makes this sort of inequality especially troublesome is that, unlike wage levels or house values, which are collected by government departments or academics, this would be a form of disadvantage that would be extremely difficult to spot.

In addition to favouring more skilled workers, digital technology also increases the financial returns to capital owners over labour. Machines don't demand a share of the profits, which means any machine-driven productivity gains accrue to whoever owns them, and that's usually the wealthy. It is of note that the percentage share of GDP going to labour relative to capital has been falling in recent years. For much of the twentieth century, the ratio of national wealth in the US between labour and capital was 66/33. It is now 58/42. The great defence against these trends for much of the twentieth century were trade unions, which ensured that the spoils of corporate profit were spread around. The unions' slow decline has been disastrous for wealth equality and – in a cruel twist – new technology is likely to further militate against worker unionisation, by both making it harder for 'gig economy' workers to band together, and by giving bosses new ways to monitor and control their workforce.* At the most extreme end of this economic bifurcation, the world's richest eight men own more than the bottom half of the world's population – and four of them are the founders of technology companies.[8]

This is not a book about the economics of digital technology – there are plenty of those already – but

* Uber and Deliveroo are part of an increasingly important sort of industry: the gig economy encompasses companies that monetise everything from borrowing cars (RelayRides), helping with daily tasks (TaskRabbit), lending bikes (Liquid) or money (Lending Club) and selling home Wi-Fi (Fon) or clothes (NeighborGoods). According to the Chartered Institute of Personnel and Development, approximately 1.3 million people are already working in the gig economy in the UK this number is predicted to grow substantially in the next few years.

about politics. While some degree of inequality is inevitable and necessary in a free market economy, too much is bad for democracy. It is well-documented that a healthy democracy depends on a vibrant, sizeable middle class. This sector of a population is democracy's backbone – the people that buy newspapers, join political parties, sponsor charities, vote and participate in community projects. Based on decades of research into the impact of inequality, we can predict with some confidence that a 'barbell economy' would result in, among other things: a shrinking tax base, growing levels of crime, depression, addiction and infant mortality, lower life expectancy and poorer health.[9]

High levels of inequality also wear away the fabric of society. The more unequal life gets, the less we spend time with people not like ourselves, and the less we trust each other. Ironically, trusting societies are more likely to be innovative and entrepreneurial, since they see the world as full of reliable people and good opportunities.[10] Most importantly of all, the middle classes are democracy's most fervent supporters. One of the reasons Marx's projected revolution didn't take place in the UK – somewhere he expected it to occur first – was because the working classes evolved into a broad, asset-owning middle class. Having something to protect and a stake in society, this group is repeatedly found in studies to value individual freedom, property rights and democratic accountability more than other groups.[11] The emergence of middle-class societies, especially in Europe and America increased the legitimacy of liberal democracy as a political system in both the nineteenth and twentieth centuries.[12]

To see what happens when tech-fuelled inequality takes off, there's no better place to start than the home of it all, Silicon Valley and its increasingly put-upon neighbour, San Francisco. There are two worlds in Silicon Valley, and they barely ever meet. There's the exciting start-up open-plan offices, with beanbags, table football, TED Talks and flip-flops, where the region's half a million tech workers can expect to earn on average well over a hundred thousand dollars a year. (For the biggest companies, the median salary is higher still.) Mostly under 40, they want to live in nearby bustling San Francisco, since Silicon Valley can resemble *The Stepford Wives.* Each morning, thousands of tech workers hop on private, Wi-Fi-enabled coaches from one of the dozens of pick-up points in San Fran's increasingly gentrified streets, and head down Highway 101 into Menlo Park (for Facebook), Sunnyvale (for Yahoo) or Mountain View (for Google). It's impossible to ignore the buzz, the thrill, and the enterprise of the place.

Alongside it, though, is another world, inhabited by the people who are left behind in the mad rush towards progress: the ignored women in tech start-ups who complain about misogyny, the Uber drivers who can only afford to live 70 miles away and have to work on zero-hour contracts, the long-time residents who are turfed out so their landlords can rent out their homes on Airbnb. It's a place where minorities struggle on low-wage service jobs, serving the largely white affluent tech workers. The median house value in both San Francisco and Silicon Valley is now around a million dollars, and the average rent is over three thousand per month for a two-bedroom flat: beyond the reach of almost anyone but tech workers. (The average salary in San Francisco is $46,000, and less if you don't work for Facebook, Google, etc.) In one of the richest cities in America, 15,000 registered homeless people – one of the highest numbers per capita in the country – struggle on, often with serious mental health and drug addiction problems. This has been a long-standing problem in San Francisco (and in much of California), though when I was there recently, locals told me that it has never been as bad as it is now. There are parts of this glorious and gleaming metropolis that reek of destitution, used needles, human waste and food banks, some of it literally in the shadows of the world's biggest and coolest companies. One morning I witnessed junkies openly shooting up on a busy pavement: it wasn't yet 9a.m. And, on the same street, techies wearing white earbuds entered the gleaming offices of a company that promises to let you 'belong anywhere'.

Epilogue: Universal Basic Income

At some point all this creative destruction becomes bad even for the winners. No one wants to live in a world comprising a handful of trillionaires and hordes of unemployed or extremely poorly paid people – not even the trillionaires. A growing number of people are proposing a bold new idea to deal with this.

In 2017 I interviewed Sam Altman, the president of Y Combinator, the most important fund in Silicon Valley for tech start-ups. Thousands of businesses apply every year to access Y Combinator's funding and guidance, in exchange for a small slice of their company. Sam is a Princeton dropout and frequently wears a hoodie, yet when I met him, he was only 31 years old and already a multi-millionaire. He is often described as 'the man who invents the future'. The companies Y Combinator have funded include Airbnb and Starsky Robotics, and are now altogether valued at $80 billion.

Aware of the potential turbulence that AI might unleash, Y Combinator recently started to fund a pilot in universal basic income. UBI, as it is commonly referred to, is an increasingly popular idea to deal with the possible rise of joblessness and tech-fuelled inequality. The basic concept is that governments should give everyone enough money to live on, with no strings attached. Several pilot schemes, including Oakland, California and Finland, are examining the idea (although it's too early to say how well they are working yet), and a number of serious thinkers and writers believe it is worth further investigation. In other words, UBI has become very fashionable. For some people on the political right, it is a way to keep capitalism ticking over in times of economic uncertainty. For some on the left, including a handful of radicals in the UK circling Labour leader Jeremy Corbyn, it represents a way to redistribute wealth more fairly. And for the utopians, it would allow people to do more meaningful things with their lives than monotonous labour.*

---

* Some of them are no doubt thinking of Karl Marx's vision of a communist paradise, where people could 'hunt in the morning, fish in the afternoon, rear cattle in the evening, criticise after dinner'.

Sam doesn't think anyone is ready for AI. 'We are going to need to have new distribution, new social safety nets,' he told me in his Y Combinator office. 'What happens if you just give people money to live on? ... Say, "Here's enough money to have a house and eat and have fun".'

It's an interesting idea. There are an awful lot of jobs that people don't really want to do. If the things that people gain from work – economic means, structure, purpose – can be achieved in other ways, that's worth exploring. Proponents of UBI argue that this would be a 'basic' income – and not necessarily a replacement for work. Some people would work while others would choose not to, and could instead dedicate their time to other things.

However, I don't see how UBI would stop a tiny band of elites from becoming even wealthier than everyone else. And who would pay for UBI is not clear to me, especially at a time when the tech firms seem keen to pay as little tax as possible. (In fact, the issue of how to pay for UBI is a quite interesting problem. If you were to divide up all current US spending on social welfare by capita, it would amount to only $2,300 per person per year, which is clearly not enough. Advocates for UBI rely on assumptions about the falling future costs of essential goods. Sam Altman, for example, in a 2016 discussion on the subject, said that it would be affordable in the future because of huge increases in productivity and a reduction in the cost of necessities. I doubt this would be a strong enough basis to persuade most policy-makers in government). 'It strains credulity,' writes tech critic Nicholas Carr, 'to imagine today's technology moguls, with their libertarian leanings and impatience with government, agreeing to the kind of vast wealth distribution scheme that would be necessary.'[13] Certainly their behaviour to date does not bode well.

I asked Sam what felt to me like a very simple question: would people really be happy living in a society in which there are a small number of very rich people, and everyone else is given money to keep them occupied. What about dignity in work? What about growing inequality?

'You have a very pessimistic view of the future,' he replied. 'I hope you're wrong. I believe that someone doing mechanical labour is not the best fulfilment of their dreams and aspirations.'

Neither do I, of course, but that's not really the point. 'The thing that makes me nervous is that society will have to change dramatically, and that's quite worrying,' I said.

'I believe society *will* have to change dramatically,' he replied. 'We've been through many of these changes before. Look, I understand that people have this spirit of "I'm going to hang on to the past at all costs [At this, Sam clenched his fists, looked up to the sky, and waved his arms] and I hate progress and I hate change", and I get that from you.'

'It's not that!' I cut back in. 'It's not hating progress. What if the progress that you are creating is not what other people want?'

'There are 40 million people in the US that live in poverty,' he said. 'If technology can eliminate human suffering, we should do that; if technology can generate more wealth and we can figure out how to distribute it better, we should do that.' There was no hint that tech has played some role in creating the problem that tech is now supposed to fix. I mentioned that journalists have to ask about the negative possibilities. That's our job.

'If you continue this thrust of "we should stop progress", no one is going to take you seriously,' he replied. 'I think you can add an important voice, but I worry you are going in the wrong direction with this anti-progress angle.'

With that, the interview was over. Sam is considered one of Silicon Valley's most interesting and original thinkers, and there's no doubting his sharp mind. But I suspect on the current trajectory, the more likely scenario is a relentless drive toward a more unequal economy. The winners always convince themselves that they deserve their small fortunes and that they are helping society become more connected, even as they profit from tearing it apart. The dystopia we should fear is not robots with all the jobs, but a barbell-shaped economy where socially progressive tech millionaires live in gated communities well away from the masses who they either fear, patronise or detest. The feeling would, of course, be mutual.

Chapter 7: The Everything Monopoly

How the Tech Giants are Taking Over the World

Chapter 5: The Everything Monopoly

How the Tech Giants are Taking Over the World

How technology unexpectedly tends to result in monopolies. Tech firms are already transferring their economic power into political power through lobbying, but they differ from 'traditional' monopolies in important ways: by owning the platforms on which material is published, they have an important influence over public opinion and activism itself. This has important ramifications for how citizens practise 'free association', which is the basis of all independent civil society and a bulwark against tyranny. On our current path we are moving into the final phase of these monopolies – of not just economics or politics, but of culture and ideas.

THE MOST EXTREME – and in the coming years probably the most pressing – illustration of how digital tech drives inequality is the creation of massive tech monopolies. This problem is related, though slightly different, to the issues discussed in the last chapter:

the tendency of powerful companies to distort politics by virtue of their size and power.

Before explaining why this is such a problem for democracies, we need a primer on how and why three of the current top five people in the Forbes Rich List are tech titans, and why the five biggest companies by market value in the world are West Coast tech firms. Lots of markets tend to concentrate on a few winners: think pharma, oil or even supermarkets. Students graduate from business schools with schemes to build monopolies, because, once established, the lack of competition allows monopolies to raise prices and increase profits. But the scale and size of today's monopolies is entirely novel.* Back in the 1990s many predicted that the internet would slay monopolies, not create them. The popular thinking of the time – repeated over and over by the era's digital gurus and futurists – was that the net was decentralised and connected, and so would automatically lead to a competitive and distributed

* The following operate as either monopolies or oligopolies in their respective fields: Google (search engine, video streaming, online advertising); Facebook (social network, messaging, online advertising); Uber (ride-sharing); Airbnb (home-sharing); Amazon (online retail – especially books, cloud computing); Twitter (micro-blogging); Instagram (photo-sharing); Spotify (music streaming).

marketplace.[1] No one knew exactly how, but influential figures like Chris Anderson called this 'the long tail' and were extremely excited about it.

It seems obvious now that the nature of digital technology makes monopolies more rather than less likely. The most important reason for this is the effect of networks. If you join Facebook, your friends will be more likely to join too, which in turn makes their friends more likely to join. When everything is connected, such network effects can spread further, and far more quickly. It is such a powerful force that the biggest problem faced by Facebook today is that it's running out of humans to connect. The same thing happens with online markets. When I was young, I bought my music at a nearby record shop, where my choices were constrained by geography and limited information. As a result I bought some niche albums. In small, local markets, there can be lots of best taxi services, best bookstores or best record shops. But, with digital markets, you only need one. Why would I use the *quite* good local taxi firm when I can use brilliant Uber?

Humans are bad at understanding the power of these network effects, because we tend to think in linear form, whereas networks can grow exponentially. And so we are stunned every time a new billion-dollar mega-corporation appears almost overnight. Network effects are so powerful because they are self-reinforcing. The more users Uber has, the more drivers (and data) it attracts, and the better service it can offer, which means it gets more users, which means it continues to grow. Google goes through this iterative nano-improvement millions of times each day and is starting to look like what economists call a natural monopoly, because it can provide a better service than two competitive firms could.[2]* Digital companies can also scale up any advantage at breakneck speed because the cost to them of expanding is often very low. It costs hardly anything for Airbnb to add a new unit, whereas for a hotel company constructing a new building is slow, expensive and risky. The cost to YouTube of hosting one video or one million is roughly the same, which can't be said of the dozen or so Blockbusters which the internet informs me stagger on, renting physical DVDs.

All this means that the best provider can capture entire industries more easily, creating winner-takes-all sectors. Strictly speaking, 'winner-takes-most' would be more accurate; although I use

* People who talk breezily of 'breaking-up' Google don't appreciate how much worse search engines would become without it.

the term 'monopoly', which means 'single seller', 'oligopoly' is the correct term for the status of the tech giants, referring to a market shared by a small number of producers or sellers. Even Google is not a monopoly in internet search engines – there are others, like DuckDuckGo or Bing. Still, rather than the hoped-for long tail of successful small vendors selling to niche markets, we instead have a few massive winners and little else. On iTunes, for example, 0.00001 per cent of tracks accounts for a sixth of all sales, while the bottom 94 per cent sell fewer than one hundred copies each.[3] I suppose that's *technically* a long tail, but it's a very thin one.*

---

* You may have heard excited talk of the latest technologies that promises to liberate us from monopolies and concentrations of power, collectively called 'blockchain'. The most famous application of this is the cryptocurrency bitcoin. These are certainly very interesting, and we'll examine them in detail in the next chapter. However, advocates of this new technology sound a lot like the techno-optimists of the nineties, again promising with huge conviction a world of peer-to-peer, distributed exchanges and the long tail. But similar patterns as happened then seem to be emerging already. A small number of people own a disproportionately large amount of the bitcoins. And because bitcoin 'mining' is reliant on having the best technology and most powerful computing rigs, the mining function of many crypto-coins are concentrated in the hands of a small number of already wealthy people and venture capitalists.

Everyone in Silicon Valley knows all this, of course. They talk about the benefits of free markets, while at the same time betting the venture capital on monopolies. Peter Thiel – the founder of PayPal and probably the most influential of all the Silicon Valley tech investors – says he only puts money in companies that have monopoly potential. Some tech firms run at short-term loss, kept afloat by venture capital on a 'growth before profit' philosophy, as they chase market domination. Uber has been running billion-dollar losses for years: when there are no competitors left, their tempting prices may well be inflated.

Economic theory suggests that markets re-adjust to monopolies – competitors up their game and newcomers try to join the action. Maybe the tech giants' success will in turn be disrupted by the arrival of feisty upstarts. There are certainly occasional anti-monopoly success stories and examples of market response (local taxi firms do seem to be getting better, in order to compete with Uber), so perhaps it's too soon to know. But once monopolies are established, they do everything they can to keep themselves there. The largest tech firms are able to recruit all the best talent, by offering wallet-busting salaries, healthcare, private buses, housing and so on. I recently visited GCHQ as part of an outreach effort held by the intelligence agency, who are worried about losing their best computer programmers to the tech firms, who can outbid even them (imagine how much worse it must be for local councils). GCHQ has a security-cleared Costa Coffee in their building with notoriously long queues and average drinks. Facebook's Menlo Park has excellent coffee.

The biggest tech firms are motoring ahead. They spend more on research than businesses in other industries: the top companies in the US that spend the most on research and development are 'the big five': Amazon, Alphabet (Google's holding company), Intel, Microsoft and Apple. And, if anyone does threaten to compete with them, they have enough cash reserve to simply buy them out before their position is challenged.* I meet lots of young start-up founders in London, and many of them are hoping to get bought out by Google or Facebook. The result is that the big boys risk squelching innovation and competition, squeezing out smaller companies and ideas. Monopolistic power also means that the biggest firms can pressure their smaller

---

* Facebook bought WhatsApp, Instagram, Oculus VR and attempted to buy Snap Inc, developers of Snapchat. Meanwhile, Amazon bought e-commerce website Zappos, Wholefoods and Audible.

rivals, especially when they own the platform on which they depend. Amazon's domination isn't of books per se – it's of the place where books are sold. That means it can set the prices and dictate terms, and other vendors have to accept it.

Like other big businesses before them, tech firms are now converting their economic power into political influence. Businesses spend millions of dollars every year in London, Brussels and Washington on obtaining influence and interests: meeting with ministers, pushing ideas on policy wonks, hosting nice lunches and the rest. The tech giants stayed out of the messy business of politics for many years, but as they've grown, so have their opportunities for influence. Big Tech spending on direct lobbying now matches that of most other industries, in both the US and the EU.[4] Google spent more than any other company on lobbying in Washington, DC in 2017 – around $18m – and all the other tech firms increased their spending too.[5]

Influence is a lot more than just hard cash. There is also a revolving door of well-educated, enthusiastic people between government and the tech industry. According to the Google Transparency Project, 53 people worked at both Google and the White House during the Obama administration. It's a similar story in the UK, with 28 people moving in either direction between UK politics and Google in recent years, including former advisers to Tony Blair and former deputy Prime Minister Nick Clegg. Five people even moved from government to Google and then back again.

There is no evidence of malpractice on the part of these companies or people, and all the ones I've met are professional and competent, which explains why they are in demand. But the result of all this movement is that policy-makers and tech firms tend to be comprised of similar people with shared views and assumptions. They attend the same events and go to the same parties. I'm sure in the back of their mind is the possibility that they might end up working for 'the other side' one day.

The ongoing digitisation of the economy means more and more aspects of our economy will be subject to this drift towards monopolies. For example, 'smart manufacturing' is the process in which every aspect of a production line collects data and communicates with every other part of the process, allowing comprehensive monitoring and analytics, in real-time. This doesn't end at the factory, either: once the product is out in the real world, embedded sensors will constantly collect data too – in smart fridges, smart toys, smart food packaging and so on. Astonishingly, this is already happening, although only in a limited way. When everything is online, different devices will need to speak to each other. Your phone will talk to your fridge, which will talk to the supermarket, which will talk to its suppliers and manufacturers: data chit-chat all the way up and down the chain. It's far more efficient if only one or two companies provide the infrastructure to do this, in the same way that Google is better when there's only one. This explains why Siemens has spent $4 billion acquiring smart manufacturing capabilities to build its industrial platform MindSphere, and why General Electric is working to build its own platform, called 'Predix'. 'It's winner takes all,' General Electric's chief digital officer said recently.

The same rule will apply to many of the AI techniques I discussed in the last chapter. AI is what's known as a 'general purpose' technology, meaning it can be applied in a wide variety of contexts. Although the specific application is very different, driverless vehicles like Stefan's Starsky trucks use similar techniques of data extraction and analysis as AI-powered crime-prediction technology or CV analysis. Google's DeepMind, for example, doesn't just win at Go – it is currently pioneering exciting new medical research and has already dramatically cut the energy bills at Google's huge data centres by using deep learning to optimise the air conditioning systems.[6] There are countervailing tendencies, of course – some experts have got together to develop 'open source' AI which is more transparent and, hopefully, carefully designed, but the direction of progress is clear – just follow the money. Over the past few years, big tech firms have bought promising AI start-ups by the truckload. Google's DeepMind is one of only a dozen they have recently acquired. Apple splashed out $200 million for Turi, a machine learning start-up, in 2016, and Intel has invested over $1 billion in AI companies over the past couple of years.[7]

Market leaders in AI like Google, with the data, the geniuses, the experience and the computing power, won't be limited to just search and information retrieval. They will also be able to leap ahead in almost anything where AI is important: logistics, driverless cars, medical research, television, factory production, city planning, agriculture, energy use, storage, clerical work, education and who knows what else. Amazon is already a retailer, marketing platform, delivery and logistics network, payment system, credit lender, auction house, book publisher, TV production company, fashion designer and cloud computing provider.[8] What next? Here's my prediction: in the next decade or so, a small number of tech firms will get an edge in AI and smart manufacturing and create the biggest cross-industry monopolies that have ever existed. At some terrible point, these tech giants could become so important to the health and well-being of the nation that they are, like large banks, too big to fail. Armed with the best tech and the most skilled engineers, maybe Google or Facebook could be the only ones who could solve sophisticated cyber-crime (perhaps committed by a powerful AI from a hostile country?), fix computer bugs, predict and pre-empt economic shocks, run the National Grid or protect the cyber defences of the big banks – cyber security in the public sector is predictably understaffed and under-skilled.[9] In the occasional discussions I have had with law makers on these subjects, I've sensed that they'd quite like to smash a tech monopoly or two, but realise it would be extremely damaging for the economy, and are therefore stuck.

Soft power

You'd be right to think that this has happened before. In some ways tech firms are simply following the examples set by dubious role models, including rail-road barons, 1980s pro-market think tanks, PR firms and oil giants. Whenever there is such concentration of economic strength, there is usually a corrosion of politics, because wealthy and powerful people will always wish to maintain and increase their power. Many years ago, Associate Supreme Court Justice Louis Brandeis said that 'we can either have democracy in this country or we can have great wealth concentrated in the hands of a few, but we can't have both'. Some degree of economic concentration is inevitable in a free market democracy, but at a certain level it corrodes and distorts the political process, because narrow business interests are always looked after at the expense of other groups.

If tech monopolies carry on their seemingly unstoppable growth and continue to transfer their economic power into political influence (and the entire history of modern capitalism strongly suggests that they will), groups that are less well-positioned will switch off from politics entirely, creating a downward spiral in which politics will descend into a game of rich companies and politicians discussing positions and ideas among themselves. As we've seen with the election of Trump and the result of the EU referendum, the idea that an 'out of touch elite' or 'professional establishment' runs politics can produce an extremely powerful and visceral response from the electorate. The thing that most people forget about these surprising votes is that the people were correct to think there is an economic and political elite of shared interests, because there is.

Over the years, most democracies have developed anti-trust legislation to guard against this. But, especially in the US, such legislation is designed around the idea that monopolies are bad for the public only when prices start going up or consumer welfare is damaged.[10] Today's tech firms are different beasts, because they often push prices *down* and are generally excellent for consumers. Some – like Facebook and Google – are technically free to use. Furthermore, even the definitions of these companies are blurred. Standard Oil was an oil company. What is Facebook? A media outlet? An online advertiser? A social media platform? An AI company?

We need to reconsider what modern monopolies look like using non-price metrics. The threat tech monopolies pose to democracies is about more than the prices they charge: it's the concentration of power, data and control over the public space – and their ability to wield this power over a growing number of economic activities, especially in the infrastructure and technologies of the future. Integrated into everything, everywhere, their technology will blanket the world. When viewed with all this in mind, the tech monopolies start to look more invidious to politics than even a monopoly from America's past like Rockefeller or Carnegie, in several important ways.

Firstly, political parties are dependent on tech firms in ways that they are not on other companies. Every politician, with only a few exceptions, values the support of business. But politicians need tech platforms to reach voters in a manner that they don't need other businesses, and these companies own and run the platforms on which so much of our political debate occurs. We should not forget what desperate people our politicians usually are. Tired, time poor, technically illiterate, when in whizzes a modern-day shaman, with millions of data points, tailored messaging and huge audiences. As I showed in Chapter Three, Facebook was embedded within Trump's campaign. Similarly, Eric Schmidt, Executive Chairman of Google, worked on Obama's re-election campaign in 2012.*[11]

Secondly, owning the infrastructure of these massive digital platforms also gives these companies the unprecedented opportunity to tweak and nudge

---

* OK, this wouldn't be completely unique either. In the nineteenth century, Western Union used to give elected officials unlimited free use of their telegraph system, which they believed was the 'cheapest means' of calming critics in Washington.

the public debate in subtle-but-beneficial ways. In September 2017, Transport for London decided not to renew Uber's license to operate in the city. According to TfL, Uber's conduct demonstrated 'a lack of corporate responsibility in relation to a number of issues which have potential public safety and security implications'. Then a strange thing happened. Uber started a petition on the website change.org and encouraged its massive user base to get involved:

> To defend the livelihoods of 40,000 drivers – and the consumer choice of millions of Londoners – sign this petition asking to reverse the decision to ban Uber in London.

Thousands of bus-shy Londoners rushed to sign it, making it the fastest-growing petition in the UK in 2017.[12] Large PR firms have in the past manufactured so-called 'AstroTurf' grassroots activism, but I don't ever recall receiving emails or phone calls from BP asking me to lobby or campaign on fuel duties, and nor has the automated checkout machine at Morrisons ever told me that, in addition to moving the unexpected item from the bagging area, I might like to tell my MP that I'd like corporation tax lowered. These traditional companies have no way of getting to me, but I carry Uber about in my pocket everywhere. They sent emails and notifications to regular users – telling them that TfL's decision 'will deprive you of the choice of a convenient way of getting about town'. Tellingly, Uber's terms and conditions were even updated recently to allow the app to 'inform you about elections, ballots, referenda and other political processes that relate to our services'.[13] (Uber's appeal against TfL's decision is ongoing.)

That a small clique of private companies have so much power over the structure and content of public debate, what information we receive and how we communicate seems to me completely and utterly insane. Tech firms understand how precious and controversial this power is, and so wave the magic wand sparingly. Back in 2012, US Congress was presented with the Stop Online Piracy Act. The film and music industry broadly backed the proposal, which was aimed at cracking down on piracy by targeting sites that linked to web pages containing illegal content. However, Google strongly opposed the legislation, and used its status as the front page of the internet to let this be known. For 24 hours visitors to the site found a giant black box over the Google logo and a link: 'Tell Congress – please don't censor the web'.[14] On clicking, it redirected to a petition that urged Congress to reject the bill. No company has ever been able to reach more people more quickly. Millions clicked the link, of course, overwhelming the web servers of Congress. The bill failed. I happen to think that not censoring the web is an honourable goal, but it is also in Google's commercial interests. The home-sharing app Airbnb has taken this a step further, spending millions of dollars on creating a community – it's always that word *community*, so warm and harmless – of grassroots activists called 'home-sharing clubs' that will be willing to fight local regulations and represent 'a powerful people-to-people political advocacy bloc'.[15] Like the Stop Online Piracy Act killed by Google, this 'community' helped defeat a piece of legislation in 2015 which would have involved changes to short-term lets.

These are merely the cases that I am familiar with. It's impossible to know what other small nips and tucks are made to drive or influence public debate in a way that suits their interests – and that, of course, is part of the problem.

The potential distortion of the independent public sphere is more important than it first appears, because it illustrates a broader point: that civil society itself is becoming arranged around platforms and abstractions rather than alert citizens practised in locally rooted action. The nightmare scenario would involve infantilised consumer-citizens hooked on and desperate for cheap and accessible goods and services, with no questions or strings attached – millions of convenience addicts, ready to mobilise at a moment's notice if an update alert tells them to.

Democracy theorists from Aristotle onwards have realised the importance of healthy and independent citizen-led bodies – charities, conservation groups, sports teams and so on – that are neither business nor government. This is because citizens getting together freely to do things is the way they are educated to be 'free and independent actors' rather than supplicants of the state or corporate interests. This is the healthy idea of the citizen – the one who actually reads the terms and conditions, who is politically alert, who thinks about workers' rights, taxation and zero hours contracts and is aware of the consequences of his or her purchasing decisions. These citizens are ready to mobilise too, but when they do, the decision will be based on their own self-interest.

Alexis de Tocqueville, who perhaps thought more than anyone else has about the role of a citizen in a democracy, wrote in *Democracy in America* that private associations are 'schools for democracy'. Autocrats and dictators – who read more about democratic theory than democrats – always dismantle independent civic society, including those that have nothing to do with politics. All tyrants realise that organised independent civil society groups carry somewhere within them the spirit of rebellion and agitation.

In February 2017 Mark Zuckerberg published a 'manifesto' for how to create a better world. Its well-intentioned thrust was that Facebook wants to bring us closer together and build 'a global community'. This, unfortunately, is a contradiction in terms. Community – meaningful community, not abstract groups of a billion virtually connected avatars – is local. There are plenty of examples of citizens using social media to enable more physical real-world activism – and that is welcome. But, as psychologist Sherry Turkle points out in her book of the same title, we are often 'alone, together', especially when online, chatting without forming meaningful bonds or learning the art of uniting for a common purpose. Physical networks of organised people practised in locally rooted action can, if necessary, stand up to governments or others that would oppress them. Millions of clickers and swipers, all connected through flimsy digital associations with others a thousand miles away, are no match for the physical power of organised bureaucracies.*

Finally, the great tech monopolies have prospered at the expense of journalism, the traditional 'fourth estate' and one of the few institutions that can shine a light on them. The story of its decline – and especially of local newspapers – has been told often enough, but the basics are these: print advertising and sales revenue is down, because fewer people read the physical paper. Online advertisers are more interested in volume than quality, and so the link between advertising spends and quality output is fractured. And because there is now so much content online, the ad spend per article is tiny. In addition, many people now view their articles through Facebook rather than on, say, a news homepage, and so much of the revenue and customer data stays with the platform rather than the media company. In some countries it's not just

* The debate on this so-called 'slacktivism' – how political groups use the internet for activism – is interesting and highly contentious. Zeynep Tufekci, perhaps the world's leading academic on how civil society groups mobilise online, argues that digital technology enables small groups to mobilise quickly and easily – but often at the cost of enabling them to make a real-world impact.

money either – social media platforms drive a high proportion of traffic to sites, which creates a relationship of dependency.[16]

I don't want to be too nostalgic about this. It is hard to measure the decline of a whole industry. While traditional media revenue (especially that of local newspapers) has fallen sharply, there are some signs of a recovery, at least for certain quality newspapers and magazines that employ subscription models.* Plus the 'old media' still wield some influence. For many years, Rupert Murdoch was something of a kingmaker in British politics, terrifying MPs and prime ministers that were desperate for the support of *The Sun* and *The Times* newspapers. His loosened grip on British politics – note how little difference his newspapers' relentless attacks in 2017 made on Corbyn's popularity – is good news for democracy. However, most journalists, imperfect though they are, do have a commitment to 'the best available version of the truth' and to holding the powerful to

* Nevertheless, The Poynter Institute estimates that Facebook has taken around $1 billion from print adverts for US papers, while Alan Rusbridger, the former editor of the *Guardian*, reckons Facebook took around £20 million from the paper's projected ad revenue in 2016.

account. The industry's decline is desperately worrying because almost every story that uncovers shadowy influences in our politics – lobbying, manipulation, corruption – is the result of painstaking, expensive journalism: the Pentagon Papers, the Edward Snowden leaks, the Paradise Papers or even the *Observer*'s recent investigations into the role played by data analytics firms during the EU referendum.

Traditional media is not dead just yet – and journalists have an important job of building public trust in their work too, which has also been in sharp decline over the past several years. But if the tech gets more complicated, more pervasive and more political (and it will) then we will need ever-more-careful – and probably extremely expensive – investigations to uncover what's going on. We need our best and most fearless minds to tackle questions like: how are algorithms run, and in whose interest? What new injustices is tech creating? Where is political influence being applied at the local level? Which AI firms and new capabilities are Google buying, and why? And what might they do with them? These are extremely difficult questions to answer. Some – like the mysterious ad tech infrastructure, surely one of the most-powerful and least-understood aspects of online life – are extremely technical.* We are therefore in the unenviable position where the tech companies could become even less subject to the investigations that could keep their growing power in check.

Economic to political to cultural

I could stop here, but the next and perhaps final stage is when economic power morphs into what Marxists sometimes call 'cultural hegemony': where domination can be achieved through controlling the ideas and assumptions available to the public. The idea – associated with the Italian Marxist Antonio Gramsci and his criticism of capitalism – is worth considering here, because there is little doubt that a techno-utopian view of the world has infected society.

---

* Advertising in the tech world works like this: there are 'demand side' platforms that decide which online ad space to buy and for how much (mostly done via some complicated big data analysis) and 'supply side' platforms which sell the space to them. The two are matched up on exchange sites, which run constant real-time auctions. Whenever you refresh a web page, an auction is conducted over who will offer the highest amount to show you an advert – this is surely one of the strangest things about the internet. Very few people fully understand how it all fits together, including the advertisers who are paying for it and the regulators who are struggling to get to grips with it.

All technology encodes within it certain values and assumptions about how the world works. Gutenberg's press was more than a mere printing machine – it popularised the ideal of free information exchange. Similarly, the nineteenth-century penny press papers created a new demand for gossip and a hard criticism of power. The telegraph system transformed people's perceptions of time and distance, while the radio helped invent the concept of a single shared nationality, culture and language. The medium, remember, is the message. And the medium of digital technology, as a sector, is now monopolising the whole economy.

Back in 1995, in an extremely perceptive paper, left-wing academics Richard Barbrook and Andy Cameron detailed the philosophy and ideas of the new tech wunderkinds, which they christened 'The Californian Ideology', a fusion of the cultural bohemianism of San Francisco and entrepreneurial free market zeal. This ideology, they thought, was appealing because it offered a way out of the traditional political struggles over wealth distribution or fairness. A profound faith in the emancipatory qualities of technology allowed the techies to paper over any inconsistencies between the yuppy and hippie ideals, because they promised that when the revolution arrived everyone would be great and cool and fulfilled and rich. All you needed to get to utopia was a belief in 'disruption', the idea that progress is achieved through smashing up old industries and institutions and replacing them with something new and digital. Steve Jobs – at once the acid-dropping hippy and the ruthless businessman – was this Californian Ideology incarnate.

This is the secret behind the digital revolution. The reason that start-ups flock to Silicon Valley is not just the promise of building a better world – it's because that's where the venture capital is. Money and ideas in Silicon Valley have a very complicated relationship. Even start-up visionaries and wide-eyed socially minded inventors need money to survive, to pay extortionate Bay Area rent and to hire the best programmers. Silicon Valley runs according to a Faustian pact: money in exchange for world-changing ideas. But investment brings with it new responsibilities, and suddenly there are profit margins, quarterlies and growth targets. In some ways, tech is just the latest vehicle for very rich people to use well-tested techniques of buying political influence, monopolistic behaviour and regulation avoidance, to help them become even richer. Doing it through tech allows them to add a glossy veneer of progress on top of some very familiar behaviour.

Over the years, the big tech firms have very carefully cultivated the Californian Ideology: even though they are massive multi-billion-dollar corporations with huge PR teams, they pitch themselves as anti-establishment; even though they are built on a model of data extraction and surveillance capitalism, they purport to be promoting exciting and liberating technology; even though they are dominated by rich white guys, they talk of social justice and equality. I sometimes think it must be very confusing to be Mark Zuckerberg. In 2014, only 2 per cent of Facebook staff were black and less than a third were women. They were also caught providing inaccurate information about user data matching to the European Commission during their acquisition of WhatsApp. And yet, later that year, Zuckerberg said that 'our philosophy is that we care about people first'.[17] The worse these companies behave and the richer they become, the more they spend on looking cool and talking about fairness and community. This cannot be a coincidence.

Wealthy corporations cultivate the popular ideas of the day not just by direct pressure, but also by funnelling money – through think tanks, TED Talks, grants, sponsorships and consulting – towards individuals and ideas that see the world as they do.[18] And through their funding of think tanks and increasingly academia, the broad public imagination about technology is rebalanced in a subtle-but-definite way.[19]

But it's much more than that. The iPhone and web browsers we now all use have carried the Californian Ideology around the world, infecting us all with the alluring idea that disruption is liberation, total individualism is empowerment and gadgets equal progress. Sometimes these things are true, though they are hardly iron laws of social change. But believing it all means the tech firms march off into the future and then come back and hand us a map to guide us through it. It is hard to imagine the coming years without schools full of iPads (Apple), VR headsets (the Facebook-owned Oculus) and coding classes (run by Google). Recent research from the NSPCC found that almost half of all children want to pursue a career in tech. An even more depressing statistic is that 30 per cent hope to become the one-in-a-million YouTuber who actually makes a career of it. Every country wants to build their own Silicon Valley, and every city has ambitions to be a tech 'hub'. Read any political manifesto from across the spectrum and you'll find yourself lost in a world of smart cities, lean governments and flexible workers.

To seriously criticise any of this puts you at risk of being labelled a Luddite who doesn't 'get it'.

And to whom do we look in order to solve our collective social problems? It's no longer the state, but the modern tech-geek superhero. Space travel and climate change has fallen to Elon Musk. We look to Google to solve health problems and sort out ageing. Facebook gets to decide what free speech is and battle against fake news, while Amazon's Jeff Bezos saves the *Washington Post* from bankruptcy and funds scholarships. One UK MP recently suggested we might run the National Health Service like Uber, while another pitched the idea of Airbnb-style room rentals for patients who needed to stay overnight. Heaven help us all.

Total victory for the monopoly is not over economics or politics – it's over assumptions, ideas and possible futures. Because when that happens, Big Tech won't need to lobby or buy out competitors. They will have so insinuated themselves in our lives and minds, that we won't be able to imagine a world without them.

Chapter 6: Crypto-Anarchy

Does Total Liberty Lead to the End of the State?

The rise of crypto-anarchy – a philosophy that aims to undermine the power of the state via encryption – is on the rise. It is driven by a quest to protect our privacy online. However, it also challenges the fundamental authority of the state, and threatens to weaken it to the point of near collapse.

A COUPLE OF YEARS ago I was invited to Prague to give a talk at the 'Institute of Cryptoanarchy'. Pavol, a mildly secretive hacker who uses several pseudonyms, was organising a gathering of programmers, libertarians and crypto-anarchists. The theme, according to the programme that he emailed over, was *decentralised.* 'The concept of the authoritative state is gradually becoming obsolete,' it read. 'The rise of sharing economies with reputation models, digital contracts and cryptocurrencies makes the role of central governments useless.'

The Institute of Cryptoanarchy is housed in a large three-storey building in central Prague called 'Parralel Polis'. It was set up in 2014 by a handful of artists and cryptography enthusiasts who wanted to explore ways of using technology to carve out more space for individual freedom. In 1968 Prague was the scene of the 'Spring' attempt by citizens to wrest freedoms from the Soviet Union. Today's authoritarians, Pavol told me, are the so-called democracies of the world who offer an illusion of liberty while controlling everything and everyone. The weekend-long event was dedicated to speeding up its demise using bitcoin, secretive messaging apps and anonymous web-browsers.

Crypto-anarchy is one of the very few genuinely original – and utterly revolutionary – political philosophies of the last 50 years. It's the politics of dystopian sci-fi movies, a mix of 'crypto' (the maths of keeping certain things hidden) and 'anarchy' (the absence of government).

I arrived at the building on Dělnická Street early one cold Saturday morning. It was easy to spot: its coal black brick façade stood out from its grey neighbours like an unusual stone on the beach, and 'Institute of Cryptoanarchy' was written in bright white letters on the front. I was a little late, and the place was already teeming with scores of men in their twenties and thirties speaking in the mid-Atlantic English that almost every hacker, crypto-enthusiast and bitcoiner seems to possess. A 3D printer whirred in the background, and bitcoin t-shirts and posters of Edward Snowden were available for sale. Within a cable's reach of every plug socket, eyes stared at lines and lines of the incomprehensible language of computers: Java, Ruby, C++. As I wandered about looking for somewhere to charge up my devices, I spotted 'The Crypto-Anarchist Manifesto' printed out and pinned to the wall:

> A specter is haunting the modern world, the specter of crypto anarchy. Computer technology is on the verge of providing the ability for individuals and groups to communicate and interact with each other in a totally anonymous manner ... These developments will alter completely the nature of government regulation, the ability to tax and control economic interactions, the ability to keep information secret, and will even alter the nature of trust and reputation.

This manifesto of techno-freedom was written by a young Californian called Timothy C. May in the late 1980s. May had made some brilliant breakthroughs in computer memory chip design at Intel, but his real interest was in how this weird new internet would change politics. He worked with mathematician Eric Hughes and computer scientist John Gilmore (internet history buffs will know Gilmore as the creator of the notorious alt Usenet group) to pursue this theme. All three were radical libertarians from California and early adopters of computer technology. While many West Coast liberals were giving well-received talks about the coming age of digital liberation, this more technically informed trio realised that digital technology would more likely create a dystopia of ubiquitous state espionage and control. They believed that this nightmare could only be averted if people had access to powerful encryption that would allow them to protect their online identities. Encryption is the art and science of keeping secrets from people you don't want to know them, while revealing them to those you do. During the nineties, as more people got online, the authorities tried to prevent people from accessing strong encryption, because they worried about cyberspace becoming a safe haven for terrorists and criminals. One type of encryption technique known as 'public key encryption' which was invented in 1976, especially bothered the FBI because it made encryption safer and far easier to use.[1]

The trio, however, wanted as many people as possible to have access to this encryption. They set up an email list and held meetings with a few dozen like-minded people to build and send it out into the world. A journalist who attended some of their early gatherings called the group 'the cypherpunks', a play on the word cypher and the cyberpunk genre of fiction popularised by sci-fi writers such as William Gibson. Rather than engage in a political battle which they would lose, these cypherpunks instead decided to build tech that would make digital spaces ungovernable by law. The first post to their growing email list was a 1987 speech given by mathematician Chuck Hammill called *From Crossbows to Cryptography: Thwarting the State via Technology.* 'For a fraction of the investment in time, money and effort I might expend in trying to convince the state to abolish wiretapping and all forms of censorship,' wrote Hammill, 'I can teach every libertarian who's interested how to use cryptography to abolish them unilaterally.'

These cypherpunks reasoned that encryption could do more than just protect citizens: they believed it could also carve out vast new spaces of freedom online and maybe push society closer to an anarchic paradise where governments would be severely weakened. Many of them believed that too many decisions that affected the liberty of the individual were determined by the vote of democratic governments. 'Politics has never given anyone lasting freedom, and it never will,' May wrote in 1993. But perhaps, he thought, technology could.

Sometimes relatively innocuous inventions open up new possibilities for social organisation. In mid-nineteenth-century America the settlement of farming communities in the West was impossible, because roaming cattle kept destroying the crops. But the invention of 'barbed wire' (sharp metal barbs twisted around a strand of smooth wire, with a second intertwined piece of wire so that the barbs couldn't slide) meant that huge swathes of land could be enclosed. Roaming buffalo were doomed, which in turn destroyed the Native American way of life (understandably, they nicknamed barbed wire 'the devil's rope').[2] Technology, according to the famous professor of technology history, Melvin Kranzberg, is 'neither good nor bad; but nor is it neutral'.

(Public key) encryption is the crypto-anarchist's barbed wire. It allows people to communicate, browse and transact beyond the reach of government, making it significantly harder for the state to control information, and subsequently, its citizens. This is because of a simple-but-magic rule: due to some arcane properties of prime numbers, it takes far more computing power to decrypt something than to encrypt it.[3] It's like an egg: a lot easier to crack than to put back in its shell. Julian Assange, who was an active contributor to Timothy May's email list, puts it this way: 'the universe believes in encryption'.

Crypto-anarchy in the UK

Over the course of the nineties this remarkable group predicted, developed or helped spread several techniques that are now routinely employed by computer users to protect against various forms of surveillance. Timothy May proposed, among other things, secure cryptocurrencies, a tool enabling people to browse the web anonymously, an unregulated marketplace which he called 'BlackNet', where anything could be bought or sold without being tracked and a system of anonymous whistleblowing. By the turn of the millennium, the US authorities had more or less given up on trying to put the software back in the box, and it looked like the crypto-anarchists had won. After 9/11 however, the threat of terrorism resulted in sweeping new surveillance powers being granted to government. And as online commerce and social media proliferated, millions of people started trading their internet privacy for free services – something the crypto crowd hadn't bargained for.

But crypto-anarchy is now back, driven by a growing realisation among ordinary users that every click we make online is being collected: by GCHQ, by Facebook, Russian hackers and who knows who else. The 500 or so people who'd turned up to the Institute of Cryptoanarchy for the weekend represent a small part of a punk movement which is trying to rekindle the dream. In fact, the return of crypto-anarchy is in some sense a direct response to the trends I've been discussing in this book. Motivated by an honourable desire to protect online freedom and privacy, hundreds of people are working on ingenious ways of keeping online secrets, preventing censorship and fighting against centralised control. Over the last couple of years there's been a flurry of 'crypto-parties' all around the world, where internet users can learn about the latest techniques to protect their privacy online. Anonymous web browsers like Tor, which can browse the net without giving away the user's location (and are used to access the 'dark net', an encrypted network of sites that uses a non-standard protocol), are becoming ever more popular. There are now hundreds of encrypted messaging apps: Signal,

WhatsApp, FrozenChat, ChatSecure, Wickr and more. WikiLeaks continues to cause mayhem by exposing state and political secrets.

The most popular crypto-anarchy technology at the moment is probably bitcoin. In case you are not familiar with it, bitcoin is a digital currency. I won't describe in detail how it works here – there are plenty of other good guides available – but here's the short version: a quantity of bitcoin is stored at a bitcoin address, the key to which is a unique string of letters and numbers that can be kept on a website, desktop, mobile phone or even a piece of paper. Anyone can download a bitcoin wallet on to their computer, buy bitcoin with traditional currency from a currency exchange, and use them to buy or sell a growing number of products or services as easily as sending an email. Transactions are secure, fast and free, with no central authority controlling value or supply, and no middlemen taking a slice. You don't even have to give your real name to start up an account. As with most new technologies, bitcoin has had teething problems,* and has been subject to

---

* Chief among them: the centralisation of mining, transaction speed, environmental costs, questionable 'initial coin offerings' and the fact that a small number of people own a high proportion of them.

wild speculation and price volatility. But these are early days. Bitcoin may not become the dominant cryptocurrency when the dust settles, but they are here to stay because of the benefits that they offer ordinary people and businesses.

What matters most, however, is that people are using these systems even though they are not backed by any central government. On arriving at the Cryptoanarchist Institute, I joined the queue for food and coffee. But my Czech currency, koruna, which I had exchanged at the airport at an extortionate rate, were not accepted. 'We only take bitcoin,' said the assistant. (I later learned that this was the one place in the world that accepted only bitcoin.) Ever since we abandoned the gold standard, all national currencies have run on trust. We accept sterling or dollars because we believe others will. And people trust bitcoin and the maths that underpins it. At the institute's cafe the staff were paid in bitcoin; rent collected for their co-working space was paid in bitcoin, too. I was given a little plastic card with a QR code, and transferred bitcoin on to it using one of three yellow ATM machines. From that point on, every time I wanted anything I just scanned the QR code. A coffee. Ping! A Red Bull. Ping! Some goulash. Ping! A postcard of Edward Snowden. Ping! I didn't use my koruna once.*

Bitcoin is more than just money, though: it's a new way of handling information. Bear with me on this short-but-important technical detour. Every time someone sends a bitcoin payment to a recipient, a record of the transaction is stored in something called the blockchain, a huge database of every bitcoin transaction ever made. Transactions are collected into blocks, with each block representing about ten minutes' worth of transactions. The blocks are ordered chronologically, with each including a digital signature (a 'hash') of the previous block, which administers the ordering and guarantees that a new block can join the chain if it starts where the preceding one finishes. A copy of the blockchain record is independently maintained by the thousands of computers which have installed the software. As a result, historic transactions cannot be undone or edited, because that would require editing every independent record. If you've ever read that bitcoin

* Given a bitcoin was worth around £300 back then, and is now trading at over £5,000, my cup of coffee cost approximately £75 in today's money. Some of the staff have probably now retired.

is 'anonymous', that's not strictly true, because of this database record. However, even though the blockchain records the transactions, there is no link to the identity of the people behind them, which is why some writers prefer to call it 'pseudonymous'. A simple way to put it is: the blockchain is a massive, distributed, tamper-proof database that anyone can add to but no one can delete.

Bitcoin's blockchain was designed to store financial transactions, but it can hold other information, too. In fact, a new wave of blockchains allow complicated code to be stored. This could be as revolutionary as the internet itself, because it represents a way to store information in a far more decentralised way. Crypto-anarchists are in raptures over this. There are t-shirts. Its top thinkers are like rock stars. The leaders of the movement hold expensive conferences in exclusive city centre venues, and meet-up groups gather every day of the week in pubs. The whole scene is bursting with zeal, energy and billions of dollars. All sorts of blockchain applications have recently been released: OpenBazaar, a peer-to-peer marketplace that is impossible to shut down, decentralised file storage, a distributed web domain name system, land ownership records in India to combat fraud and prediction markets. Several are working on social media applications that are impossible to censor or control because they're hosted on a decentralised blockchain.

Perhaps the most important functionality of the new wave of blockchains is the way they allow 'smart contracts', lines of code that execute instructions automatically. They were first proposed by Nick Szabo (yet another cypherpunk who was on the original mailing list) way back in 1994.[4] It works like this: you set up a contract in the form of a programme that is triggered when a condition is met – for example, a payment when an invoice is filed, and which cannot be interfered with once it's deployed.

Crypto-anarchy is not on the point of taking over society. As an explicit philosophy and movement it remains on the fringes, and most people who use encryption would not consider themselves crypto-anarchists. After all, the technology already underpins most of the world's e-commerce transactions, as well as popular messaging apps like WhatsApp. Crypto tools and ideas have already entered the mainstream, but where's that taking us?

The joys of anonymity

These technologies help democracy in several important ways, and they certainly advance the cause of individual freedom. I'm fortunate to be writing in the luxury of a stable democracy, but in vast swathes of the world, restrictive governments deny their citizens anything like freedom of conscience or expression. Encryption helps protect individuals – including journalists, who are of course vital in a democracy.

Equally there are myriad positive uses for bitcoin (and the hundreds of other cryptocurrencies that work according to similar principles). Being able to instantly send money to anywhere in the world with no fees, charges or banks will be especially liberating for people in countries with an over-leveraged banking sector run by corrupt politicians. It might even provide a secure digital payment option for the millions who are still excluded from the formal banking system. These are not trivial benefits. The economic boon of blockchain is potentially staggering – especially if twinned with the internet of things. Imagine a bridge with embedded sensors which could detect minor faults and necessary repairs. It could also track which vehicles have used it. Once a threshold of faults is reached, a smart contract could be automatically initiated, with every user charged immediately proportionate to their use. This could even have big benefits for how government works. The British Government hopes these immutable databases will create opportunities for a 'greater transparency of transactions between government agencies and citizens'. Estonia, meanwhile, is charging full steam ahead. Citizens there can already log into their own health records stored on a blockchain using a digital identity to see if any medical professionals have looked at their data – and if they have, they can demand to know why.

Crypto-anarchy might even be the only set of ideas that can challenge the tech monopolies that I examined in the previous chapter, and the death of free will I discussed in Chapter One. People like Timothy May spotted the dangers of total connectivity 20 years before the rest of us. Many of the problems I have covered in this book are a restatement of crypto-anarchists' fears about surveillance and data-driven managerialism. Some of their techniques will be vital in the coming years to defend ourselves against the tech giants' ceaseless pursuit of dataism.

However, the end-point of this revolution is more far-reaching than its supporters realise. In the well-intentioned pursuit of privacy and freedom, we might risk undermining the entire edifice on which these rights are based. Most liberals have been very short-sighted about this, because they want total freedom *and* equality, without realising that the two are sometimes in tension. This is why the issue of encryption and privacy throws up peculiar political alliances. (The most notable of recent years is surely the idiotic social democratic love affair with crypto-anarchist Julian Assange.)

Democracy is about individual liberty of course, but that's only half the picture. It is also a system of coercion because your liberty must sometimes be taken away too. The state must be able to force you to pay tax, remove your passport, restrict your right to assemble and back it up with the use of force if it needs to by arresting you and throwing you in prison. The state's control of information justifies and organises this system of coercion, through official taxation records, land registries, criminal records, censuses and passports. The moral basis for this control is the assertion that its laws and powers express the will of the people, and also protect certain fundamental rights.

Crypto-anarchy is dynamite against state control because it challenges the government's authority to coerce people within its borders and to control information. The crypto-anarchists believe our rights and freedoms should not be reliant on democratically passed law, but rather on immutable technology that no human law, no judge, no police officer, can change or crack. The crypto-anarchist says information shouldn't be held on some secretive database controlled by government, but on decentralised systems controlled by no one. Rousseau wrote that on birth we enter into a social contract where society grants us rights but also assigns us responsibilities, which the state is able to enforce by violence. Crypto-anarchists want to opt out of the latter, but that will come at the cost of the former. By challenging the state's authority at such a fundamental level, the spectre of crypto-anarchy now hangs over the core functions of democracy, just as Timothy May predicted.

Law & disorder

Let's take law and order, often thought of as the first duty of the state. Even without crypto-anarchy, the internet is a huge problem for the police. It magnifies both our creative and destructive faculties. This is a good thing for individual freedom but a bad thing for law enforcement agencies, who find their scope of work increasing all the time – and who are often helpless to respond. The more connected we are, the more vulnerable we are. A Russian can now steal your money without leaving his bunker in Volgograd. If I were so inclined (I'm not) I could turn on my anonymous Tor browser, jump onto the dark net, fire some ransomware into the world, and wait for bitcoin ransom payments from the unsuspecting internet users who had clicked on my malicious link. None of this requires much in the way of skill or know-how.[5]

And yet successful prosecution for cybercrime is negligible. There's barely a thing our police can do about Russian hackers. They cannot stop the trade in stolen data. They're struggling to remove illegal pornography from the internet. I'm not suggesting that crypto-anarchists are happy about this – they aren't. And for individual citizens, better encryption is one solution to the problem of digital crime. However, the more crypto-anarchy spreads, the worse this will become. There are already plenty of crypto-anarchist corners of the internet where the king's writ barely runs. Silk Road, an anonymous dark net marketplace, was set up a couple of years after the invention of bitcoin. It used the anonymous Tor browser to obscure the location of buyers and sellers, encrypted messaging for communication and bitcoin as a means of payment. Between 2011 and 2013 it processed over $1.2 billion worth of sales, mostly illegal drugs. Although Silk Road was eventually shut down, there are now several other dark net markets, where stolen personal data, narcotics and child abuse images can be bought and sold, Amazon-style.

Let's imagine a blockchain-based social media platform (there are already versions of this, like

Mastodon, which is on the normal net), in which posts are simultaneously hosted on multiple decentralised blockchain databases. Facebook runs on servers that sit in massive data centres controlled by the company – meaning that it can delete or edit what its users see. A blockchain social media platform would be untouchable – no government would be able to edit or remove hate-speech, illegal images or terror propaganda, unless the whole network was somehow vaporised. Blockchain advocates hate 'middle men'. They talk a lot about using tech to get rid of them, advocating contracts without managers, invoices without accountants, banks without bankers. But sometimes middle men are useful. The police often complain that Facebook and Twitter are slow or unresponsive at executing legal take-down requests. However, both these companies follow the law in whatever land they are in – and they can do that because they own the servers. One day soon, the police will look back fondly on these firms, remembering a time when there was at least a middle man they could lean on. With decentralised networks, you might as well pass laws to change the orbit of the moon.[6]

Advocates don't realise how much this risks undermining public confidence in the police and the criminal justice system as a whole. What happens when, for example, our police simply cannot remove illegal material from the web? Or when they cannot prosecute cybercriminals or stop malicious software? Part of the 'deal' of living under a democracy is that we give up certain individual liberties in order to secure other collective rights, most obviously security. The rise of crypto-anarchy means that governments might soon struggle to hold up their end of this bargain.

The reason this is so important is because I suspect future technology will increase further the ability of small groups of individuals to do great harm, which means the authorities will need *greater* power, not less. For reasons still not entirely clear to me, humanity is currently embarked on a quixotic quest to connect everything to everything else. Within a decade, your TV, dog, house, car, fridge and clothing, will be part of the invisible internet of things network, all chipped and communicating with each other. Sometimes they will be lifesaving: a smart fire alarm might immediately turn on your phone alarm, unlock your door and contact the fire brigade. But they will also be vulnerable, because the security standards for these 'IoT' devices are notoriously bad. There have already been high-profile examples of cardiac devices, cars, a baby monitor and home webcams being hacked. This will get very personal. It won't be long, for example, before your smart coffee machine will be hacked with ransomware – and you are asked to pay a small ransom just to regain access to your morning caffeine.

Every day it gets a little simpler to be a cybercriminal. Earlier this year it was reported that there is now easily available code called AutoSploit that automatically searches for vulnerable IoT devices. Once it finds them, it scans the Metasploit database, which lists hacking exploits, to find the best form of attack. This is fully automated crime: you set the programme running and then it disappears into cyberspace and hacks whatever it can. In addition to difficult legal questions (what exactly could you be prosecuted for, if you hadn't known which devices would be attacked?) it means you need zero skills to become a hacker. Tools like this are usually designed for security experts to identify weaknesses and fix them, which is a noble endeavour, but they are also frequently exploited by those with less honourable intentions. What happens if you pair this open, decentralised system with AI? Marc Goodman in his recent book *Future Crimes* speculated about a powerful AI system controlled by a modern-day Al Capone that carries out hits by hacking cars and deliberately crashing them. It's not just an online problem either. AI expert Stuart Russell has proposed one especially terrifying scenario: widely available bumblebee-sized drones that could kill using minimal explosive power by shooting people in the eye. One day, he reckons, these will be affordable and could be programmed remotely to fly to targets, identify them using facial recognition, and kill them, before self-destructing to avoid the possibility of any diagnostics. And once the government bans these terrifying flying death drones, they will re-emerge on dark net markets that the authorities would be unable to shut down.

None of this will happen overnight, so don't bother locking your windows. And the police aren't helpless – even people on the dark net or those using cryptocurrencies sometimes get caught. If the authorities really decide to come after you, there's usually a way. The problem, however, is that it's getting far more expensive and time consuming to find and prosecute online criminals, which means that the police do less and less of it. And as the cost of law enforcement is going up, the barriers to entry into criminality are going down.

When it comes to crime, there is always an arms race between attack and defence, and ever since the establishment of modern policing and criminal justice, the two have been in something like a manageable balance. But it is foolish to assume that they always will be, and the evidence of the last few years is not promising.

If you think this is worrying – and I believe you should – then the challenge to government posed by bitcoin is potentially existential, because it is a direct challenge to the state monopoly over money. Many within the bitcoin community these days think of it as a currency that is more efficient and less subject to political whims, rather than an all-powerful black-market tool to overthrow the system. But if money becomes independent, governments will struggle to pay for themselves. Central banks have the right to print paper money, which means they can increase or reduce the amount printed (one way of raising revenue) and more easily monitor where it flows, which helps them prosecute irregularities or raise taxation. Bitcoin, however, is a medium of exchange and a store of value that is international, pseudonymous and not controllable by government. 'We had money backed by political command,' writes crypto-specialist Dominic Frisby, 'now we have money backed by mathematical proof.'

Its technical design gives some clue as to why this might be. Before bitcoin, crypto-anarchists had for years been dreaming of decentralised anonymous payment systems. The cypherpunk email list discussed it frequently. When the list wound down at around the turn of the millennium, one of the members, Perry Metzger, set up a new cryptography forum to carry on these discussions. In late 2008, someone called Satoshi Nakamoto (in keeping with the crypto-anarchists' love of anonymity, to date no one knows who he is) first posted his idea for bitcoin. Nakamoto distrusted the global banking system, and imagined bitcoin as a way to undermine it. He hated that bankers and governments held the key to the money supply and could manipulate it to their own ends. He placed a cap on the number of bitcoins that could ever be produced (21 million) and a timetable for how quickly they could be created. This was to make sure that no central governments or central banks could print more to inflate the economy for political purposes. Although bitcoins can be bought and sold with real-world currency, new ones are not minted. Instead, anyone who dedicates his or her computing power to verifying the transactions in the public ledger blockchain competes to earn a very small amount of new bitcoins (this is called 'mining'). Its peer-to-peer, encrypted and quasi-anonymous system was designed to make linking a bitcoin transaction to a real-world person difficult, thereby making collecting taxes and monitoring users awkward.

I don't think our current banking system is perfect, but if these cryptocurrencies really take off it will create a lot of problems. First, it would be harder for government to raise income tax at source. At the very least it would mean an increase in self-assessment, which usually means a diminished tax return, partly due to confusion and difficulty with monitoring.[7] There would surely be an increase in tax evasion too, along with a rise in money laundering, since it could be done using cryptocurrencies without the hassle of legal business fronts or official bank accounts.[8] (Bitcoin has a public ledger but other cryptocurrencies, like Monero and Dash are even harder to follow.) Who knows? Some businesses might put themselves entirely on a blockchain, and be paid entirely with untraceable cryptocurrency.

Authorities in various countries have started to look more closely into how to tax cryptocurrencies – although at the moment this has mostly concerned capital gains rather than income or spending taxes.[9] If history is anything to go by, the most crypto-savvy will be able to engage in all sorts of spectacular and untraceable tax evasion, and a greater burden will fall on an increasingly aggrieved and angry middle class. Given the scale of the social challenges we already face – regarding healthcare, environmental change, crime, welfare – diminishing a democracy's tax-raising powers at this point is not a smart move. It is worth remembering that the English, French and American Revolutions all began in some sense because of disagreement about the conditions under which tax could be legitimately raised. And just as there should be no taxation without representation, so there can be no representation without taxation, because there would be no money to provide services. In fact, what would be the point of having representatives at all if there was no money to spend?

One way to understand the fundamental nature of the crypto-anarchist challenge to the state is to consider how the authorities are responding. There is exactly zero chance that governments of the world will give up on taxation or censorship – they will try to crush crypto-anarchy first. When the founder of Silk Road Ross Ulbricht, also known by the pseudonym 'Dread Pirate Roberts', was finally caught, he was sentenced to multiple life imprisonment without the possibility of parole. On delivering this draconian sentence in 2015, Judge Forrest told the court that Silk Road's very existence was '... deeply troubling, terribly misguided, and very dangerous'. The reason he was given such an exemplary punishment is because a functioning anonymous online marketplace is a direct threat to the state's authority. It's the same story with the punishments handed down to 'Anonymous' hacktivists by US courts, who routinely receive jail time for vandalising websites. In the UK, the Home Secretary has proposed increasing the maximum sentence of viewing extremist material online to *15 years in prison*, a suggestion worthy of an authoritarian regime. Consider too the sorts of punishment meted out to whistleblowers like Chelsea Manning. (My suspicion is that bitcoin is next.) Such absurd penalties for online crime can only be understood as a sign that governments are beginning to recognise how serious this is, and that deterrence might be their only remaining weapon. This is not a sign of their strength, of course, but of their weakness.

There is another, deeper reason that crypto-anarchy is on the rise. The crypto-anarchist has an almost faith-like belief in the power of technology over politics. He looks around the world (and back over history) and sees a lengthy charge sheet of oppression, corruption and suffering caused by democratic political decisions. Let's be honest, democracy hasn't so far done a particularly good job at regulating financial markets. Despite some regulation of the sector following the recent global financial crisis, bankers' bonuses are still astronomical and many of those responsible for the crash have carried on like nothing happened. Democracies have annihilitated the environment and seem incapable of responding. Many democracies are unduly influenced by donors and lobbyists, which skew politics away from the ordinary people. Billions of dollars each year vanish into offshore accounts and complicated constellations of shell companies. So it's no surprise that people are losing faith. A recent survey in the *Journal of Democracy* found that only 30 per cent of US millennials (the demographic made up of those born since 1980) agree that 'it's essential to live in a democracy', compared to 75 per cent of those born in the 1930s, and results in most other democracies demonstrate a similar pattern.[10] The failures of democracy make crypto-anarchy a more attractive proposition, especially to the young. Compared to democracy's erring and stuttering, it has the same appeal as a religious theocracy – immutable, perfect and eternal.

Lots of privacy activists believe in encryption for very honourable and educated reasons. Most have no intention of creating a stateless crypto-anarchic utopia. But that is where we might end up. Timothy May, perhaps the single most influential person behind crypto-anarchy, believes that in the coming decades, democracies as we know them will disintegrate. I managed to interview May a couple of years back. He is delighted at recent events. 'Man, this has got to be freaking big brother out!' he told me. 'We're about to see the burn-off of useless eaters,' he added, only half-joking. 'Approximately four to five billion people on our planet are essentially doomed: crypto is about making the world safe for the one per cent.' That's the final realisation of the crypto-anarchist fantasy. A world of lonely one-per-centers, freed from all constraints and social commitment – anonymous ghosts in the machine.

Conclusion: Say Hello to the Future

There are competing ideas about how society might be transformed by technology. More likely than a tech-fuelled age of freedom is actually the reverse: that growing numbers of people will turn to authoritarian ideas and leaders to restore control and order to society. Will democracy be destroyed slowly, under the guise of saving it?

I'VE TAKEN YOU THROUGH the ways each of my six pillars of democracy are being weakened, but no one knows for sure how the current tech revolution will play out. I've found that there are two broad political scenarios that most people have in mind, which I'll call the 'utopian' and 'dystopian' visions. Which is which, of course, depends on your own political viewpoint.

A growing number of people from both the left and the right of politics imagine that the falling cost of goods and higher machine-driven productivity will produce a world of plenty and the end of meaningless work. Our lives will be happier, easier and more fulfilling. Greater connectivity and more information will continue to make us generally wiser, better informed and hopefully kinder. But, to make sure people aren't left behind, something akin to a universal basic income will be needed to spread the wealth around. For many people this is the utopian scenario.

By contrast, the dystopian scenario is that central governments will gradually lose the ability to function properly. Inequality will increase to a point where a tiny number of people end up with all the tech and all the wealth and everyone else has no choice but to scratch out a living serving the winners. Governments will lose their authority, power and right to represent. And, as order slowly breaks down, the richest will disappear into heavily defended forts, as in Ayn Rand's *Atlas Shrugged.* One man's nightmare in another's fantasy: for the hard-line crypto-anarchists like Timothy May, this is a necessary and welcome step on the path to a post-state paradise of cryptocurrencies and borderless virtual communities.

Obviously there are more nuanced and realistic shades of these two scenarios. I am not a futurist, but I believe both underestimate how much will be dictated, not by the tech itself, but by how the winners and losers respond to the changes tech will unleash.* So here's a third possibility, based on what might happen if all the trends I've identified in this book unfold simultaneously. It won't be quite the way I paint it of course (the world is too unpredictable), but I think it's roughly right. It's a warning rather than a roadmap: if we can imagine a future, perhaps we can also figure out how to avert it.

Growing inequality, which I think seems unavoidable at this point, would worsen many social problems, including depression, alcoholism and crime. As Pickett and Wilkinson argue in *The Spirit Level*, greater economic inequality in a country results in more big government because the demand for police, healthcare, prisons and social services all go up. And yet simultaneously the tax base would be falling, due to that unholy alliance of the gig economy, offshore monopolies and cryptocurrencies. This means governments will be increasingly unable to do what citizens ask of them in response to these challenges and risks what the social scientist Francis Fukuyama calls a 'low level equilibrium', where poor quality government breeds

---

* Examples of excellent books on technology which do not really consider the way in which technological advances and changes will be shaped by politics include Max Tegmark's *Life 3.0* and Nick Bostrom's *Superintelligence*. The opposite is true as well: Steven Levitsky and Daniel Ziblatt's *How Democracies Die* barely mentions technology at all.

the distrust of citizens, who then withhold the compliance and resources necessary for government to function effectively. It is a self-reinforcing problem, and I wonder if we aren't already witnessing this destructive equilibrium.

The social side effect of growing inequality would be an increasingly fractured society comprised of different social and ethnic groups whose jobs, schools or paths never cross, online or off. One predictable new fault line of inequality could be between the tech-haves who enjoy the benefits of personal AI bots, high productivity and stupendous healthcare, and a less savvy underclass. The former would become more engaged and the latter even less so, leaving politics more open to capture from a tech elite along the lines I suggested in Chapter Five about monopolies.

If this came to pass, large numbers of people would start to see machines as forces of control and repression rather than of liberation.* Are there not

---

* Once questions of freedom and the human spirit are invoked, it's impossible to know how far opposition can go. Between 1978 and 1995 the 'Unabomber', aka Ted Kaczynski, sent 16 bombs to targets that included universities and airlines, killing three people and injuring 23. Kaczynski, a Harvard maths prodigy who had disappeared to live off-grid in his twenties, was motivated by a belief that technological change

already signs that people are turning against technology? Witness the rise in 'the digital detox', off-grid communities and anti-Uber protests over the past few years. Imagine what might happen when driverless cars and Starsky trucks turn up – does anyone seriously think that drivers will passively let this happen, consoled by the fact that their great-great-great grandchildren will probably be richer and less likely to die in a car crash? And what about when Trump's promised jobs don't materialise, because of automation?

This is not democracy collapsing, but rather straining at the seams – high levels of inequality, social division, economic hardship and a weak and incompetent government. This wouldn't lead to either the utopia or dystopia that I described, but rather seems like dangerous circumstances for democracy to dip toward some new flavour of

---

was destroying human civilisation and would usher in a period of dehumanised tyranny and control. He set out his ideas in a 30-thousand-word anti-tech manifesto entitled *Industrial Society and Its Future.* Once you get past Kaczynski's casual racism and calls for violent revolution, his writings on digital technology now seem uncomfortably prescient. He predicts super-intelligent machines dictating society, the psychological ill effect of overreliance on technology and huge inequality in a world run by a techno-savvy elite.

authoritarianism. According to Steven Levitsky and Daniel Ziblatt, academics who study democratic failure, we imagine democracies to end at the hands of men with guns, but they can also go down quietly via elected governments who slowly remove the institutional handrails, with the support of polarised, divided and angry citizens. In my scenario the real threat is that large numbers of people conclude that democratic values and institutions no longer solve social problems, reduce crime or create jobs.

So what might people turn to instead? It's easy to imagine a growing taste for a more 'system one' demagogue who promises to restore order, control and stability – even at the cost of undermining democratic institutions and norms.* We should be very worried, for example, that the World Values Survey has found increasing support for authoritarian leaders over the past 20 years across many democratic nations.[1] I doubt however that millions of

* The Polish parliament recently voted to hand the ruling Law and Justice Party control of judicial appointments and the Supreme Court – under the guise of speeding up the process and breaking the grip of what it calls the 'privileged caste' of legal experts. Over in Hungary, the Fidesz Party has been chipping away at the independent press for several years. It's not manipulative demagogues behind this: the people want it.

people will suddenly rush en masse to vote for a fascist or a neo-Leninist. As David Runciman argues in his forthcoming book, *How Democracy Ends*, we should not keep looking back to the 1930s for clues. (For one thing, the median age in Weimar Germany was 25, and in most democracies today it is around 20 years older. Fascism is a young man's game.)[2] Democracy today will surely end in new and different ways, and it won't happen overnight – this is a long-term problem.

One possibility is a new enthusiasm among governments and citizens for increasingly authoritarian anti-democratic techo-solutions to the problems democracy seems ill-equipped to handle. After all, powerful autocrats armed with even more powerful tech could probably make people wealthier and more gadget-rich. They could use their brilliant AIs to solve climate change, tackle out-of-control crime, energy problems, hunger and other problems that are coming down the line. Automated decision-making machines could more efficiently allocate resources without the involvement of irrational, ill-informed people. Life expectancy would shoot up if everyone's health data was analysed by powerful, secretive algorithms. Government-run cryptocurrencies with zero privacy would be insanely efficient and tax receipts would soar.

In the hands of a techno-authoritarian, all the digital tools of liberation could easily become powerful tools of subtle coercion that might make society run more smoothly but wouldn't make us more free or hold the powerful to account.

This will unfold in different ways depending on the specific area, but let's take law and order. As I mentioned in the last chapter, the growing democratisation of crime will force governments to become more draconian just to hold the ring and keep crime at manageable levels. Cash-strapped police forces will be forced to rely more on big data and crime-prediction software, which will be a cheap and efficient way to keep crime to socially acceptable levels.* But the development of big data and predictive policing techniques will exacerbate existing biases and inequalities rather than dealing with the underlying issues.

Just imagine the possibility that all the clever techniques developed by Michal Kosinksi and others to peddle jeans could be redeployed to predict our movements and our propensity to be gay or radical or criminal or critical or whatever else the authorities don't like – and to shift us off course even

* This already exists, most famously PredPol and CompStat.

before we do anything wrong. The Chinese Government is currently working up a Social Credit System to rate the trustworthiness of its 1.3 billion citizens, which they say will enhance trust and 'sincerity' through absolute transparency and monitoring. It will be like a credit rating that scores people on every facet of their life (credit, social, personal and professional). A citizen's score will affect their odds of finding a date or a marriage partner.[3] The policy document states, of course, that this citizen score 'will forge a public opinion environment where keeping trust is glorious.'

And as for bitcoin, consider a cryptocurrency in which every single transaction ever made was recorded in a database run by the government, who hold all the passwords. All your money and transactions would be collected and analysed by a central authority, and would remain on immutable databases, linked to another blockchain that would hold all your health records, personal data and credit details.

It's not impossible that our progressive tech-elite could become the demagogues themselves, or at least ally themselves to this line of thinking. They could easily become opponents of the idea of democracy, believing that the rabble can't be trusted and falling for the allure of running society themselves.

Over the last 200 years, individual liberty and wealth have grown hand-in-hand, because freedom was good for the economy, and that economy produced more well-off people who valued freedom. What if that self-reinforcing cycle was weakened? What if economic growth in the future no longer depended on individual freedom and entrepreneurial spirit, but on capital and the ownership of smart machines that can drive research and entrepreneurship? What need would the rich then have for the poor they neither knew nor liked? In this scenario, 'universal basic income' wouldn't be a dreamy utopia of satisfied and empowered citizens, but instead a very neat way for the millionaires to keep the poorest in society from rebelling.

This is why the predicted decline of the middle class is so worrying. They are our bulwark against this dystopia. Ever since the arrival of mass participatory democracy, the middle class, rather than wealthy progressive liberals, have been its most ardent supporters. This is why the digital addiction, the erosion of free will, the barbell-shaped economy and the mobbish divisions I've warned about are so serious. If what was traditionally the middle class becomes sufficiently enfeebled by growing inequality and too addicted and dependent on machines, it will neither spot what's coming nor have the time, inclination or resources to act. After all, the risk to democracy I'm painting is more subtle than military leaders taking over the radio stations and storming parliament. A weakened independent media, a civic society compromised of tech-funded NGOs and online activists without organisational experience or resourcefulness would be no match for the slow drift towards techno-authoritarianism. The *idea* of democracy won't disappear, especially in an age where everyone has a voice and a platform. We could even still have plebiscites and MPs and the rest. But it would be little more than a shell system, where real power and authority was increasingly centralised and run by a small group of techno-wizards. Instead of great liberty, uncontrolled freedom could lead us to a new form of gentle, effective and subtle techno-authoritarianism. Many of us won't notice, and those who do mostly won't care.

The Silicon Valley preppers

Not enough people are sufficiently worried about this. Liberating gadgets and tech feel intuitively like they're good for democracy, because they're good for individual freedom. But that blinds us to the larger problems. The people who build this tech, by contrast, are starting to get very worried.

I recently visited Antonio García Martínez, who until a couple of years ago was living the techie's dream life: that of a start-up guy in Silicon Valley, surrounded by hip young millionaires in open-plan offices. He'd sold his online ad company to Twitter in 2014 for a small fortune, and was working as a senior executive at Facebook (an experience he wrote about in his bestselling book *Chaos Monkeys*). But at some point in 2015, he looked into the not-too-distant future and saw a very bleak world that was nothing like the polished utopia of connectivity and total information promised by his colleagues. 'I've seen what's coming,' he told me. 'It's very scary, I think we could have some very dark days ahead of us.' So, just passing 40, he decided he needed some form of escape. He now lives most of his life on a small island called Orcas off the coast of Washington State, on five acres of land that are only accessible by 4x4 via a bumpy dirt path through densely packed trees. Instead of gleaming glass buildings and tastefully exposed brick, his new arrangements include a tepee, a building plot, some guns and ammunition, a compost toilet, a generator, wires and solar panels.

Antonio isn't the only tech entrepreneur wondering if we're clicking our way to dystopia.

Reid Hoffman, co-founder of LinkedIn and an influential investor told the *New Yorker* in 2017 that around half of all Silicon Valley billionaires have some degree of what he called 'apocalypse insurance'. PayPal co-founder and influential venture capitalist Peter Thiel recently bought a 477-acre bolthole in New Zealand and became a Kiwi national. Others discuss survivalism tactics in secret Facebook groups: helicopters, bomb-proofing, bitcoin, gold. It's not all driven by fears about technology – terrorism, natural disasters and pandemics also feature – but much of it is. According to Antonio, many tech-entrepreneurs in Silicon Valley are just as pessimistic as he is about the future they're building – they just don't say it in public. Still, the survivalist approach of guns and tepees seemed like overkill to me. 'What do you have?' he asked, fiddling around with a tape measure outside his giant tepee. 'You're just betting that it doesn't happen.' And before I could answer, he told me precisely me what I had. 'You have hope, that's what you have. Hope. And hope is a shitty hedge.'

In ancient Greek mythology, sailors would occasionally brave a pass through the narrow Strait of Messina that separates Sicily from the Italian mainland. According to the legend, on one side was Scylla, a terrifying sea monster that devoured any ships that strayed too close. Attempting to steer clear of this would, however, take the ship close to an equally dangerous hazard on the other side of the strait – a deadly whirlpool called Charybdis. Digital technology is behind the slow unravelling of power and control in democracies. The obvious monster is Scylla – turbo-charged inequality and social breakdown. But in trying to avoid it, democracies could end up in the thrall of Charybdis, a digitally powered techno-authoritarian, and wind up with China and Russia undermining democracy in the name of order and harmony. Democracy somehow needs to hold a course, as it always has done, between the two gravitational pulls of control and freedom. That means to embrace the technology that can make us better off, healthier and more fulfilled, but also ensure that it is subject to democratic control and works in the public interest.

If Locke, Rousseau, Jefferson, Montesquieu – each in their own way architects of modern democracy – were transported to 2018, they'd be dazzled by our smart phones, planes, bitcoins, hospitals, emojis and rocket launchers. They'd also be amazed to discover that we still run our democracies in the same way as in the days of carts and horses, muskets and candles. Each phase of democracy should be a product of its time – its genius is that it can change. Like artificial intelligence, democracy is a 'general purpose' technology. Ancient Athenians could manage face-to-face city-level democracy. Once society became too large and complex, representative democracy emerged as a way to keep it working. The mass party and the mass taxation systems of democracy were then layered on top when industrialism and mass suffrage arrived. Since then it has somehow stopped evolving. While this book has examined the problems of technology, many of those failures are partly down to democracies' inability to keep up with the rapid changes taking place.

Now, however, there is a new battle looming over the best way to run society: should it be governed by technology or the people? Is democracy really still the best way to ensure wealth and stability? These are spiritual as well as technical questions. At the moment technology appears to have the answers. To stand any chance of winning, democracy must offer exciting visions in a world of big data, smart machines and ubiquitous connection, and offer a believable way to get there. In the epilogue that follows, I've laid out 20 ways that we can try to do it, though it won't be easy.

We are cursed to live in interesting times. Democracy has changed itself before, and can again. At this point, it is hard to foresee how this will end. But unless we change course, democracy will be washed away by the tech revolution and join feudalism, supreme monarchies and communism as just another political experiment that worked for a while but was unable to adapt when technology evolved, and quietly disappeared.

Epilogue: 20 Ideas to Save Democracy

DEMOCRACY WON'T SAVE ITSELF. To survive the digital age, we need a combination of drastic action from our citizens and bold ideas and radical reform from our leaders. Democracy needs to refresh itself for the digital age and regain the trust and confidence of citizens. It can start by fortifying each of its six pillars with moral authority and strength. This will be a long-term challenge with no immediate fixes. But here are 20 ideas which can help.

Alert, independent-minded citizens invested with moral autonomy

OWN YOUR OPINION

While a shortage of time and attention in our ever-accelerating culture means we are always looking for help to make our choices and decisions, whether from a comparison website or Google Maps, always beware of outsourcing the responsibility to think for yourself. What can seem helpful in the short term will enfeeble you in the long term. This is all the more dangerous when it comes to making political and moral decisions.

FIGHT DISTRACTION

Being yourself is not a given in the digital age; it takes real effort and investment to assert and defend what John Stuart Mill called 'the freedom of mind'. Think of every micro-gesture online as a political statement that can have an impact, and deserves your attention. Plan your personal time and space carefully, or you'll become a slave to internet addiction and the relentless, frenetic nature of life online – at the cost of your powers of concentration and focus. Have switch-off times, avoid the 'checking cycle' and never, *never* hit refresh. As with all addictive practices, you need to moderate yourself with real discipline. Think of it as part of your duty to be an alert citizen.

A NEW DIGITAL ETHICS

In line with ex-Google designer Tristan Harris's 'Time Well Spent' movement and his championing of 'meaningful interactions', we need to shape a new digital ethic that is shared by the tech giants and encourages them to design services that aid human well-being not just maximise clicks. There must be a firm distinction ruled between ethical and unethical persuasion. The attention economy must be replaced with an economy of human value.*

A democratic culture with a commonly agreed reality and a spirit of compromise

SMASH YOUR ECHO CHAMBER

It is very easy to blame others, but all of us have a responsibility to uphold decorum online. A helpful starting point is to make a concerted effort to *listen* to what your opponents are saying, rather than dismissing them or suspecting them of having nefarious motives. Try the 'principle of charity', which means seeking out the best possible interpretation of your opponent's view and working from there. Politics should be raucous and argumentative, but also based on the underlying belief that rivals can have reasonable differences of opinions. Make a deliberate effort to break out of your echo chamber by seeking

* More information about this initiative is available from their website. In late 2017, Mark Zuckerberg started to use the phrase 'time well spent' frequently. http://humanetech.com/

alternative information sources, joining new Facebook groups or creating different feeds. Place yourself in the position of someone unlike yourself, taking that charitable frame of mind with you. And always remember the golden rule of the internet: no one is ever as annoying in real life as they seem online.

TEACH CRITICAL THINKING

It's not all on us citizens. Our education system needs to respond to the overwhelming and confusing information world. Every school should teach the critical thinking necessary to navigate the internet sceptically. The ability to judge the merits of different pieces of information isn't new, but a specific body of skills and knowledge is now needed: a combination of 'classic' techniques (such as source verification), new knowledge about how the digital world works (such as algorithms or video splicing) and a deep understanding of our own psychological biases and irrationalities. It's not just young people who are subject to online misinformation. There are plenty of books and resources available to everyone to get savvy when it comes to life online.*

* There are lots of good online fact-checking resources, notably 'PolitiFact' in the US and 'Full Fact' in the UK. However, fact-checking is rarely enough alone – a broader approach to critical thinking is necessary.

POLICING THE ALGORITHMS

Secretly designed algorithms are already creating data-led bias and invisible injustices and we urgently need a democratic mechanism to hold them to account. Our lawmakers – whether national or international – must create accountability officials who, like IRS or Ofsted inspectors, have the right to send in technicians with the requisite skills to examine Big Tech algorithms, either as random spot-checks or in relation to a specific complaint. While it may no longer be easy to 'look under the bonnet' of modern algorithms, careful examination and oversight are still possible. This is especially true during elections, where governments must demand explanations and justifications for changes in news feeds and search results that might impact on the information the public receives.

BREAK THE AD MODEL

As they say, 'If you're not paying, you're the product'. An internet economy run on the ad-based model is turning us into data points, and this has to stop. But it only works because of our complicity. Support this change with your political voice – look for greater transparency and use services that don't collect and sell personal data (consider more paid-for premium systems), strengthen your privacy settings and download ad-blockers.

Elections that are free, fair and command public trust

UPDATE ELECTION CAMPAIGN LAWS

Analogue regulations need to be brought up to speed with the digital reality. The Electoral Commission must insist that all social media spending be recorded and shared transparently – and be prepared to investigate any misuse of personal data or spending irregularities.[1] Political parties should be required to publish databases of every data point, advert and targeting technique they use during an election. Journalists and academics can then analyse it and expose any wrongdoing. The requirement of transparency should keep campaigns (slightly more) honest, and even discourage the more sinister techniques, like psychographics.*

* There are a lot of difficult questions that need to be resolved. One is how candidates prove they've spent what they say they

CELEBRATION DAY

The more micro-targeted we are, the narrower the public commons becomes. Combatting that means creating opportunities for citizens to engage with ideas and people outside their own bubbles. Election days should be public holidays, providing an opportunity for citizens to reflect on the dazzling combination of promises, pledges, half-truths and bullshit they've been subjected to over the course of the campaign. This day (or perhaps the day before, because there are legal limits on campaigning on the day of an election itself) should include hustings, debates and meet-up groups.

BOT-WATCH

Banning them will be impossible, so someone needs to keep track of the bots, trolls and other influencers

---

have (small platforms don't always have good billing systems). A much bigger problem is how to measure and regulate the large volume of small spending by third-party organisations during an election (known as 'non-party campaigning'). Donations of money, goods, property or services worth more than £500 should be reported to the Electoral Commission, but it is extremely difficult to work this out accurately. For example: how much 'value' should be attributed to an influencer sharing a post with a million followers?

that now work to shift public opinion one way or another during elections. There are already good examples of independent groups keeping a watchful eye on this, like Oxford University's 'computational propaganda' team and the Alliance for Securing Democracy, which was established in 2017 with the aim of countering attacks on democracy, including from online manipulation. Social media platforms have started to take more responsibility for policing their sites, and wherever possible they should work with these watchdog organisations to share data, intelligence and expertise in identifying manipulation techniques and make them known to the public.

Manageable levels of equality, and a vibrant middle class with a shared investment in society

SPREAD THE WEALTH

Creating the relatively stable and strong middle classes of the last century did not happen by chance – we need the same level of imagination and intervention today to manage the transition of our economies. This will almost certainly require more government intervention in the economy. One way is heavy investment in retraining schemes and incentivising job creation, especially in emerging industries like climate change adaptation, bio-tech and the health sector. Transport for London and other local authorities should build their own versions of popular apps or products (like Uber), where the profit stays with the workers, rather than going to the venture capitalists. Governments should also invest more in building (or convening standards) for the underlying architecture of the future economy, such as the driverless cars network, which should be a publicly owned and publicly run (or at least publicly regulated) utility. This would be a platform on which car companies could compete fairly.

ROBOT TAXES

We will need new ways to raise taxation – land, resource and carbon taxation all need to be re-examined because corporation and income tax are both likely to decrease in coming years. One way is to levy a tax on robots that displace human workers. If a human was working in a factory earning £25k and is replaced directly by a robot, the owner of the machine should be taxed at broadly the same rate. Measurement and compliance will be difficult, but taxing robots would in reality just be a new type of tax on capital – with the benefit of it being raised in the jurisdiction where the robot worked, which is simpler than trying to squeeze corporation tax out of a company 'based overseas'.

NEW SAFETY NETS

As the economy changes, we need to experiment with new forms of wealth distribution and social security. While universal basic income is a popular idea at the moment I don't think it will work economically or socially. But it is worth testing further. The key to future employment is likely to be continuous learning and skills development because the labour market is likely to change quickly as new technology is adopted. People can no longer expect to leave school or university with a set of skills that will see them through to retirement. Therefore we should pilot a 'Universal Training Income', where citizens have the right to be paid by the state to retrain in certain industries.* This is a way to encourage people to keep developing professionally in order to pick up the skills needed for the future.

---

* One alternative is gifts of capital, tied to education. The Royal Society of Arts, for example, recently proposed a one-off £10,000 'capital' payment to all under 55s.

WORKERS' RIGHTS

As more jobs become precarious, the middle class will depend increasingly on secure rights and decent pay. Enforcement of minimum wages and sick pay in the gig or broader 'precarious' sector is derisory and needs to be toughened up. We also need to reverse the trend of profit accruing disproportionately to capital rather than labour. One way to do that is for governments to make it easier for members of the gig economy – drivers, cyclists, handymen – to become unionised, for example requiring gig economy companies to support a platform for their workers to organise.

A competitive economy and an independent civil society

FAIR TRADE BROWSING

We users have built the modern mega-monopolies, and our ongoing addiction to free digital services (and cheap taxis) is making them stronger. Some of the responsibility lies with people who use these apps or services. We need to be aware that cheap or free services have invisible costs: whether that's your own rights or those of the workers who are employed by the companies that run them. We need to break the monopolies through our online choices – we can stop feeding the data monsters. There are lots of smaller, ethical companies providing social media, internet search engines, taxis or home rentals. Research them and be responsible in your decisions – look out for those that have ethical data usage, share profits fairly with workers and are peer-to-peer or open source. They might be more expensive and less efficient, but we should realise this is a price worth paying.* And remember that good journalism needs to be paid for, so make a point of subscribing or donating to it. That includes local newspapers, which are both a source of local accountability and a training ground for the next generation of journalists.

ANTI-TRUST

There needs to be a re-conception of the modern monopoly, based on some combination of data, market share or cross-market holdings – price or consumer welfare are no longer sufficient metrics.

* For example, use Bandcamp over Spotify. If you can afford it, use the local taxi company over Uber; use Etsy over Amazon; use DuckDuckGo instead of Google.

Governments need to be more confident in bringing anti-trust cases to break up these new monopolies where appropriate, and in creating local regulations and tough privacy controls to prevent exploitation. The General Data Protection Regulation (GDPR) which is due to come into law across Europe shortly after this book goes to print, is a good example and must be enforced with vigour.*

SAFE AI FOR GOOD

Artificial Intelligence must not become a proprietary operating system owned and run by a single winner-takes-all company. However, we cannot fall behind in the international race to develop strong AI. Non-democracies must not get an edge on us. We should encourage the sector, but it must be subject to democratic control and, above all, tough regulation to ensure it works in the public interest and is not subject to being hacked or misused.[2] Just

---

* The GDPR is an EU regulation which strengthens individual data protection for EU citizens; for example, it requires companies to obtain more explicit consent from users about collecting and sharing their personal information. All foreign companies must abide by these regulations when processing data of EU residents. It is the most significant legislation relating to data passed by the EU.

as the inventors of the atomic bomb realised the power of their creation and so dedicated themselves to creating arms control and nuclear reactor safety, so AI inventors should take similar responsibility. Max Tegmark's research into AI safety is a good example of this.*

A sovereign authority that can enforce the people's will – but remains accountable to them

THE TRANSPARENT LEVIATHAN

Maintaining law and order in the coming years will require a significant increase in law enforcement agencies' budgets, capabilities and staff numbers. This means recruiting a new wave of 'digital police' to patrol the virtual streets and developing new forms of online intelligence, engagement and forensics. This will make civil liberties groups understandably worried. Therefore, any increases in policing powers will need a

* Max Tegmark, a prominent AI expert, is co-founder of the Future of Life Institute, a non-profit organisation which researches the challenges technology presents. One important aspect of their work – which received a major donation from Elon Musk – is a global research programme aimed at ensuring that AI is beneficial for humanity.

commensurate increase in oversight and scrutiny. For example, in the UK, security-cleared members of the public should sit on the Intelligence and Security Committee and the Independent Police Complaints Commission.

REGULATE BITCOIN

Bitcoin and blockchain technologies are exciting but currently running out of control. Where possible, cryptocurrencies should be regulated by the financial authorities (especially so-called 'initial coin offerings' and coin exchanges, where money is first raised or exchanged). They should be subject to the same regulations that currently exist to combat money laundering and the financing of terrorism. This would require coin exchanges and wallet services to report suspicious transactions where possible, and to even carry out due diligence for customers trading over certain levels.[3] The tax authorities also need to urgently update tax guidance for how to pay tax on crypto-assets, and to enforce it. This experimentation must include tax raising: blockchain databases could potentially make tax raising far smoother and more efficient. The Bank of England should create its own 'official' cryptocurrency, which could be used by vendors for fast and efficient payments – but in a way that's regulated.

FUTURE GOVERNMENT

The technologies in this book might be currently undermining democracy, but they also offer exciting opportunities to dramatically improve the way that government works. We need a bold programme of reform, which brings democracies up to speed. First, there's scope for government departments to make better and more efficient decisions using data and AI. Smart meters could help save energy bills for people, welfare payments could be better targeted and police resources better allocated – provided this is all done ethically, with public involvement and with humans in the loop. Similarly, powerful AI used in the public interest could yield remarkable benefits in health research, spending decisions, intelligence, strategy and much more. Beyond that, technology like blockchain could dramatically improve how people can hold their governments to account. Central and local governments should explore ways to use blockchains to improve the functions of democracy. For example, we're used to governments making regular spending pledges at election time, which then simply evaporate. Blockchain-based accounting and contracts could help connect pledges with actual outputs – transforming how the public can track the way our taxes are spent. The UK Government should investigate whether blockchain-based identity systems – whether for land ownership, health records or passports – could improve citizen data security and efficiency, without the government accruing too much power.[4] There are many exciting new ways to involve people more in political decision-making too, including secure online voting; but these need to be adopted cautiously – votes every week on every subject is a bad idea.

Taken together, these suggestions all amount to a defence of politics over technology. Rapid technological change can empower, liberate and enrich us, but only if it is subject to powerful democratic systems which have the authority and power to act – but are also accountable to people and the public interest. It's easy to forget when surrounded by iPhones and VR headsets what can be achieved when democratic governments embrace powerful technologies and shape them in the public interest. In July 1969, as the private sector in the US was obsessing over advertising on colour television, a human first set foot on the moon, thanks to government scientists, government research and public funding. And just three months later, to far less fanfare, another group of government-funded academics, working on a computer-sharing project, transmitted a message from the University of California Los Angeles Sigma 7 Host Computer to the SDS 940 Host Computer at the Stanford Research Institute. It was the first time two computers had remotely communicated, and the 'ARPANET' was born. A decade and a few adjustments later this government research project had a new name: the internet.

Notes

Chapter 1: The New Panopticon

1 William Davies, *The Happiness Industry* (Verso, 2015), gives a very good overview of these early days.
2 John Lanchester, 'You are the product', *London Review of Books,* 17 August 2017.
3 Elizabeth Stinson, 'Stop the Endless Scroll. Delete Social Media From Your Phone', www.wired.com, 1 October 2017.
4 Adam Alter, *Irresistible* (Penguin Press, 2017).
5 Matt Richtel, 'Are Teenagers Replacing Drugs With Smartphones?', *New York Times,* 13 March 2017.
6 Adam Alter, *Irresistible.*
7 Tristan Harris, 'How Technology is Hijacking Your Mind – from a Magician and Google Design Ethicist', www.thriveglobal.com, 18 May 2016.
8 Robert Gehl, 'A History of Like', https://thenewinquiry.com, 27 March 2013.
9 Kathy Chan, 'I like this', www.facebook.com, 10 February 2009.
10 Tom Huddleston Jnr, 'Sean Parker Wonders What Facebook Is "Doing to Our Children's Brains"', www.fortune.com, 9 November 2017.
11 Natasha Singer, 'Mapping, and Sharing, the Consumer Genome', *New York Times,* 16 June 2012.
12 Michal Kosinski, David Stillwell, and Thore Graepel (2013), 'Private traits and attributes are predictable from

digital records of human behaviour', *PNAS*, 110 (15), 5802–5805.

13 Sam Levin, 'Facebook told advertisers it can identify teens feeling "insecure" and "worthless"', *Guardian*, 1 May 2017.

14 Dave Birch, 'Where are the customer's bots?', www.medium.com, 30 December 2017.

15 Evgeny Morozov has written about this at length in his book *To Save Everything, Click Here* (Allen Lane 2013).

16 Angela Nagle, *Kill All Normies* (Zero Books, 2017).

17 'The outstanding truth about artificial intelligence supporting disaster relief', www.ifrc.org, 28 November 2016.

Franklin Wolfe, 'How Artificial Intelligence Will Revolutionize the Energy Industry', www.harvard.edu, 28 August 2017.

Alex Brokaw, 'This startup uses machine learning and satellite imagery to predict crop yields', www.theverge.com, 4 August 2016.

Maria Araujo and Daniel Davila, 'Machine learning improves oil and gas monitoring', www.talkingiotinenergy.com, 9 June 2017.

18 Cathy O'Neil, *Weapons of Math Destruction* (Penguin Books, 2016). O'Neil also has an excellent blog at www.mathbabe.org detailing similar instances.

Chapter 2: The Global Village

1 Marshall McLuhan, *The Gutenberg Galaxy: The Making of Typographic Man* (University of Toronto Press, 1962).

2 Eric Norden, 'The Playboy Interview: Marshall McLuhan', *Playboy*, March 1969.

3 James Madison, 'Federalist No. 10 – The Utility of the Union as a Safeguard Against Domestic Faction and Insurrection', 23 November 1787.

4 Thomas Hawk, 'How to unleash the wisdom of crowds', www.theconversation.com, 9 February 2016.
5 See especially the following authors: Zeynep Tufekci, Eli Pariser and Evgeny Morozov. On 'post-truth', see books by Matthew D'Ancona, James Ball and Evan Davies.
6 Bruce Drake, 'Six new findings about Millennials', www.pewresearch.org, 7 March 2014. A survey repeatedly found that millennials have fewer institutional attachments than their parents, are more politically independent, but do 'connect' to personalised networks.
7 Daniel Kahneman, *Thinking, Fast and Slow* (Farrar, Straus and Giroux, 2011).

S. Messing and S.J. Westwood (2014), 'Selective exposure in the age of social media: Endorsements trump partisan source affiliation when selecting news online'. *Communication Research*, 41(8), 1042–1063.

E. Bakshy, S. Messing and L.A. Adamic (2015), 'Exposure to ideologically diverse news and opinion on Facebook', *Science*, 348 (6239), 1130–1132.
8 Jonathan Taplin, *Move Fast and Break Things* (Macmillan, 2017).
9 Lee Drutman, 'We need political parties. But their rabid partisanship could destroy American democracy', www.vox.com, 5 September 2017.
10 Joel Busher, 'Understanding the English Defence League: living on the front line of a "clash of civilisations"', 2 December 2017, www.blogs.lse.ac.uk. See also *Responding to Populist Rhetoric: A Guide* (Counterpoint, 2015).
11 Joel Busher, *The Making of Anti-Muslim Protest: Grassroots Activism in the English Defence League* (Routledge, 2015).
12 Dratman, 'We need political parties'.
13 Kate Forrester, 'New Poll Reveals Generations Prepared To Sell Each Other Out Over Brexit, www.huffingtonpost.com,12 April 2017.

14 Jonathan Freedland, 'Post-truth politicians such as Donald Trump and Boris Johnson are no joke', *Guardian*, 13 May 2016.

Miriam Valverde, 'Pants on Fire! Trump says Clinton would let 650 million people into the U.S., in one week', 31 October 2016, www.politifact.com. Polling data was taken from www.realclearpolitics.com poll tracker.

15 B. Nyhan and J. Reifler (2010), 'When corrections fail: The persistence of political misperceptions', *Political Behavior*, 32 (2), 303–330.

16 Dolores Albarracin et al. (2017), 'Debunking: A Meta-Analysis of the Psychological Efficacy of Messages Countering Misinformation, *Psychological Science*, 28 (11), 1531–1546.

17 Paul Lewis, '"Fiction is outperforming reality": how YouTube's algorithm distorts truth', *Guardian*, 2 February 2018.

18 Nicholas Confessore, 'For Whites Sensing Decline, Donald Trump Unleashes Words of Resistance', *New York Times*, 13 July 2016.

19 Southern Poverty Law Centre, 'Richard Bertrand Spencer', https://www.splcenter.org.

Confessore, 'For Whites Sensing Decline'.

John Sides, 'Resentful white people propelled Trump to the White House – and he is rewarding their loyalty', *Washington Post*, 3 August 2017.

20 *Intimidation in Public Life*, Committee on Standards in Public Life, December 2017.

Chapter 3: Software Wars

1 Joshua Green and Sasha Issenberg, 'Inside the Trump Bunker, With Days to Go', www.bloomberg.com, 17 October 2016.

2 Theresa Hong, 'Project Alamo – How I Crossed the Line in the Sand', medium.com/@alamocitychick, 29 March 2017.

3 Ian Schwartz, 'Trump Digital Director Brad Parscale Explains Data That Led To Victory on "Kelly File"', www.realclearpolitics.com, 16 November 2016.
Theresa Hong, 'How Trump's Digital Team Broke the Mold in 2016', www.mycampaigncoach.com/, 3 August 2017.

4 Hong, 'Project Alamo', ibid.

5 Hong, 'How Trump's Digital Team Broke the Mold in 2016', www.mycampaigncoach.com/, 3 August 2017.
Issie Lapowsky, 'What did Cambridge Analytica Really do for Trump' Campaign', www.wired.com, 26 October 2017.

6 Jody Avirgan, 'A History Of Data In American Politics (Part 1): William Jennings Bryan To Barack Obama', www.fivethirtyeight.com/, 14 January 2016.

7 Frederike Kaltheuner, 'Cambridge Analytica Explained: Data and Elections', www.medium.com/privacy-international, 13 April 2017.

8 Nick Allen, 'How Hillary Clinton's digital strategy helped lead to her election defeat', www.telegraph.co.uk, 9 January 2017.
Ashley Codianni, 'Inside Hillary Clinton's Digital Operation', www.edition.cnn.com, 25 August 2015.
Shane Goldmacher, 'Hillary Clinton's "Invisible Guiding Hand"', www.politico.com, 7 September 2016.

9 James Swift, 'Interview / Alexander Nix', www.contagious.com, 28 September 2016.

10 'With up to 5,000 data points on over 230 million American voters, we build your custom target audience, then use this crucial information to engage, persuade, and motivate them to act.' https://ca-political.com/ca-advantage.

11 Joshua Green and Sasha Issenberg, ibid.
12 Sue Halpern, 'How He Used Facebook to Win', *New York Review of Books*, 8 June 2017.
13 'How Facebook ads helped elect Trump', www.cbsnews.com, 6 October 2017.
14 Robert Peston, 'Politics is now a digital arms race, and Labour is winning', *Spectator*, 18 November 2017.
15 Carole Cadwalladr, 'British courts may unlock secrets of how Trump campaign profiled US voters', www.theguardian.com, 1 October 2017.

Data Protection Act 1998, http://www.legislation.gov.uk/ukpga/1998/29/contents
16 Jim Waterson, 'Here's How Labour Ran An Under-The-Radar Dark Ads Campaign During The General Election', www.buzzfeed.com, 6 June 2017.
17 Ibid.
18 Heather Stewart, 'Labour takes to the streets and social media to reach voters', www.theguardian.com, 21 April 2017.
19 Cited in Taplin, *Move Fast and Break Things.*
20 E. Goodman, S. Labo, M. Moore and D. Tambini, (2017), 'The new political campaigning', *LSE Media Policy Project Series.*

This is something Facebook itself boasts about of course. It claims to have reached over 80% of Facebook users in marginal seats in the UK election: 'Using Facebook's targeting tools, the [Conservative] party was able to reach 80.65% of Facebook users in the key marginal seats. The party's videos were viewed 3.5 million times, while 86.9% of all ads served had social context — the all-important endorsement by a friend.'

Nina Burleigh, 'How Big Date Mines Personal Info to Craft Fake News and Manipulate Voters', www.newsweek.com, 6 August 2017.

One of Facebook's most useful services is called 'Lookalike Audiences' – it allows an advertiser to provide Facebook with a small group of known supporters, and ask Facebook to expand it out. Facebook can create groups of people who are similar to the initial group and then target them.

21 Helen Lewis, 'How Jeremy Corbyn won Facebook', www.newstatesman.com, 20 July 2016.
22 J. Baldwin-Philippi (2017), 'The myths of data-driven campaigning', *Political Communication*, 34(4), 627-633.
23 Tamsin Shaw, 'Invisible Manipulators of Your Mind', *New York Review of Books*, 20 April 2017.
24 Francis Fukuyama, *Political Order and Political Decay*, (Profile, 2014).
25 Carole Cadwalladr, 'Vote Leave donations: the dark ads, the mystery "letter" – and Brexit's online guru', www.theguardian.com, 25 November 2017.
26 Tom Hamburger, 'Cruz campaign credits psychological data and analytics for its rising success', www.washingtonpost.com, 13 December 2015.
27 Matea Gold and Frances Stead Sellers, 'After working for Trump's campaign, British data firm eyes new U.S. government contracts', www.washingtonpost.com, 17 February 2017.
28 Carole Cadwalladr, 'I made Steve Bannon's psychological warefare tool', *Observer*, 18 March 2018.
29 Nina Burleigh, ibid.
30 Lucy Handley, 'Personalized TV commercials are coming to a screen near you; US marketers to spend $3 billion on targeted ads', www.cnbc.com, 15 August 2017.
31 E. Goodman, S. Labo, M. Moore and D. Tambini, ibid.
32 Vyacheslav Poonski, 'How artificial intelligence silently took over democracy', www.weforum.org, 9 August 2017.

33 Jonathan Albright, 'Who Hacked the Election? Ad Tech did. Through "Fake News," Identity Resolution and Hyper-Personalization', www.medium.com/tow-center/, 31 July 2017.

34 Nicholas Thompson and Fred Vogelstein, 'Inside the two years that shook Facebook – and the world', *Wired*, 12 February 2018.

'How Facebook ads helped elect Trump', www.cbsnews.com, 6 October 2017.

35 Esquire Editors, 'The Untold Stories of Election Day 2016', www.esquire.com, 6 November 2017.

36 Brian Stelter, 'In their own words: The story of covering Election Night 2016', www.money.cnn.com, 5 January 2017.

37 Ben Schreckinger, 'Inside Donald Trump's Election Night War Room', www.gq.com, 7 November 2017.

38 Gregory Krieg, 'The day that changed everything: Election 2016, as it happened', www.edition.cnn.com, 8 November 2017.

39 Stelter, ibid.

40 Esquire Editors, ibid.

41 'How Facebook ads helped elect Trump' October 06 2017, www.cbsnews.com, 6 October 2017.

42 Steven Bertoni, 'Exclusive Interview: How Jared Kushner Won Trump The White House', www.forbes.com, 22 November 2016.

43 Matea Gold and Frances Stead Sellers, ibid.

44 Richard Hofstadter, 'The Paranoid Style in American Politics', www.harpers.org, November 1964.

Chapter 4: Driverless Democracy

1 Frank Levy & Richard Murnane, *The New Division of Labour* (Princeton, 2004).

2 Neural networks are complicated to understand. Another way of thinking about deep learning is that the machine is fed data and works out the rules itself. For example, with image recognition it works out which bits of a picture of a dog distinguish it as a dog. It is often impossible to tell how the machine worked out these rules – this is known as the interpretability problem.
3 Cited in Robert Peston, *WTF* (Hodder & Stoughton, 2017), p.215. Adam Smith spotted this in the eighteenth century. In *The Wealth of Nations* he predicted that machines would allow 'one man to do the work of many' and that would drive up productivity and profits. This, in turn, would stimulate the demand for more labour, because it allowed owners to hire more people and build more factories. Research from Georg Graetz and Guy Michaels has found that, while manufacturing employment has fallen in most developed countries between 1996 and 2012, it has fallen less sharply where investment in robotics has been greatest.
4 'Automation and anxiety', *The Economist*, 25 June 2016.
5 According to Martin Ford, futurist and author of the award winning book *Rise of the Robots* it won't happen immediately but within a decade or so.
6 *Stick Shift: Autonomous Vehicles, Driving Jobs, and the Future of Work*, March 2017, Centre for Global Policy Solutions.
7 Mark Fahey, 'Driverless cars will kill the most jobs in select US states', www.cnbc.com, 2 September 2016.
8 'Real wages have been falling for longest period for at least 50 years, ONS says', *Guardian*, 31 January 2014.

'The World's 8 Richest Men Are Now as Wealthy as Half the World's Population', www.fortune.com, 16 January 2017.
9 David Madland, 'Growth and the Middle Class' (Spring 2011), *Democracy Journal*, 20.

10 Richard Wilkinson & Kate Pickett, *The Spirit Level* (Penguin, 2009).
11 Wilkinson & Pickett, *The Spirit Level*, pp.272-273.
12 Fukuyama, *Political Order and Political Decay.*
13 Nicholas Carr, *The Glass Cage* (Bodley Head, 2015).

Chapter 5: The Everything Monopoly

1 Douglas Rushkoff, one of the more self-aware of these people come close to an apology for his previous work in his recent book *Throwing Rocks at the Google Bus* (Penguin, 2016).
2 Before he become Google's Chief Economist, Hal Varian wrote a book called *Information Rules* (Harvard Business Review Press, 1998), where he summed this all up very well: 'positive feedback makes the strong get stronger and the weak get weaker, leading to extreme outcomes.'
3 This, according to data available through Nielsen SoundScan, cited in *Throwing Rocks at the Google Bus* by Douglas Rushkoff.
4 Duncan Robinson, 'Google heads queue to lobby Brussels', *Financial Times*, 24 June 2015.

Tony Romm, 'Apple, Amazon and Google spent record sums to lobby Trump earlier this summer', www.recode.net, 21 July 2017,

Data is also available from the website www.google-transparencyproject.org. In 2015, Microsoft spent €4.5m on lobbying – the same figure as energy companies Shell and ExxonMobil. Google's spending went from €600,000 in 2011 to €3.5m in 2015. The giant also met 29 senior officials in the first six months of 2015 – more than any other company. By 2017 they were spending $4.2m. Between 2014 and 2017, Google increased its spending by 240 per cent, while Facebook's spending more than doubled between 2016 and 2017, to a million Euros. Both are in the top ten most influential organisations as measured by the number of high-level lobby meetings with the European Commission. Google has met with representatives from the Commission from nearly every portfolio, including agriculture and humanitarian aid. Uber has increased its lobby spending sevenfold since 2015, although from a low base."

See also Andrew Keen, *How to Fix the Future* (Atlantic, 2018), p.69.

5 Hamza Shaban, 'Google for the first time outspent every other company to influence Washington in 2017', *Washington Post*, 23 January 2018.

6 Matt Burgess, 'Google's DeepMind trains AI to cut its energy bills by 40%', www.wired.com, 20 July 2016.

7 Synced, 'Tech Giants Are Gobbling Up AI Startups', www.medium.com, 4 January 2017.

Matthew Lynley, 'Google confirms its acquisition of data science community Kaggle', www.techcrunch.com, 8 March 2017.

8 'Does Amazon Present an Anti-Trust problem?' *Financial Times* Alphachat Podcast, September 2017.

9 Farhad Manjoo, 'Google, not the government, is building the future', *New York Times*, 17 May 2017. According to the earning reports, five companies (Amazon, Apple, Facebook, Alphabet and Microsoft) are on track to spend more than $60 billion on scientific research in 2017 – roughly the same as the US federal government.

10 Jefferson especially wanted 'restriction of monopolies' in the Bill of Rights – but Hamilton, who represented the New York moneyed classes fought it. The Hamilton faction won out. President Theodore Roosevelt used the 1890 Sherman Act in his first term, which fined monopolies, to break up Rockefeller's Standard Oil Trust. He wanted to make sure massive monopolies could be

controlled to act in the public interest. He worried about 'wealthy and economically powerful men whose chief object is to hold and increase their power'.

Lanchester, 'You are the product'.

Frankin Foer, *World Without Mind* (Jonathan Cape, 2017) p.191.

11 Foer, *World Without Mind*, p. 114

12 You find the petition on the www.change.org web page under 'Save Your Uber'.

13 The Uber Privacy Policy is available on their website. https://privacy.uber.com/policy

14 This website is archived at www.web.archive.org. Search for www.google.com, and search for the date 18 January 2012.

15 Biz Carson, 'Airbnb just pulled out a clever trick to fight a proposed law in San Francisco', www.uk.businessinsider.com, 7 October 2015.

Shane Hickey and Franki Cookney, 'Airbnb faces worldwide opposition. It plans a movement to rise up in its defence', *Observer*, 29 October 2016.

Heather Kelly, 'Airbnb wants to turn hosts into "grassroots" activists', www.cnn.com, 4 November 2015.

16 Since 2015, Facebook has been the biggest driver of traffic to media sites. Some publishers – especially smaller, local ones – stake everything on Facebook, and then disappear if it changes its algorithms. In October 2017, journalists in Guatemala and Slovakia expressed fear that Facebook news feed changes would dramatically change their politics. The company ran an experiment in which professional media was removed from the main news feed in several countries and placed on a second 'explore' feed. One journalist in Guatemala said 66 per cent of their traffic disappeared overnight. Similarly, Google tweaked its algorithm to make sure – so it said – that fake news fell down its ranking. It clobbered Alternet, a site dedicated to fighting

white supremacy – their traffic collapsed, falling by 40 per cent almost overnight.

17 Steven Levy, 'Mark Zuckerberg on Facebook's Future, From Virtual Reality to Anonymity', *Wired*, 30 April 2014.

18 Andrew Wilson, 'The Ideas Industry', www.thinktheology.co.uk, 16 August 2017.

19 This is all available from the website http://googletransparencyproject.org. While there are doubtless instances where collaboration and funding from the private sector benefits academics, institutions and students, according to the Google Transparency Project, of 330 studies about policy issues directly relevant to Google's operations and revenue – subjects like anti-trust, privacy and data security, net neutrality, copyright – 54 per cent were either partly funded by, or affiliated with academics or institutions funded by Google. In the majority of cases they supported positions that would be advantageous for Google. (And very often were published at the same time as investigations into Google's practices and major legislative decisions.) In August 2017 Barry Lynn, who ran a research team inside the New America Foundation think tank, praised the decision of the European Commission to levy a large fine on Google for anti-competitive practice. The New America Foundation had received $21m from Google over the years (their main conference room is even called 'The Eric Schmidt Ideas Lab' – yuk). Anne Marie Slaughter, the Foundation's Director, told Mr Lynn that he was 'imperilling the institution as a whole', and he was forced to leave. Google denied forcing Mr Lynn out.

Chapter 6: Crypto-Anarchy

1 From the time of the Roman Empire until the 1970s, encryption was based on a 'single key' model, with the

same code both locking and unlocking the message. Modern computing made encryption far more powerful, but the underlying principle was the same: if you wanted to communicate secretly with someone, you still had to get the code to them – which presented the same problem you started with: you couldn't really trust that your communications would be genuinely private. Two MIT mathematicians called Whitfield Diffie and Martin Hellman solved this in 1976 with a system they called 'public key encryption'. Each user is given his own personal cypher system comprised of two 'keys', which are different but mathematically related to each other through their relationship to a shared prime number. The mathematics behind it is complicated, but the idea is simple. It means you can share your 'public' key with everyone, and they can use it to encrypt a message into a meaningless jumble that can be decrypted only with your secret 'private' key. The public key is mathematically derived from your private key, but using reverse mathematics to derive the private key would take the world's most powerful supercomputer many trillion years to crack. This was a revolution in cryptography. The public key can be widely distributed without compromising security; the private key, however, is private. For complicated mathematical reasons, a message encoded with either key can be decoded with the other. According to David Kahn, a prominent expert, this was 'the most revolutionary new concept in the field since … the Renaissance'.

2 Roman Mars, 'Barbed Wire's Dark, Deadly History', www.gizmodo.com, 25 March 2015.

3 Timothy May offers an explanation in *Cyphernomicon*: 'I did find a simple calculation, with "toy numbers", from Matthew Ghio: "You pick two prime numbers; for example five and seven. Multiply them together, equals 35. Now

you calculate the product of one less than each number, plus one. $(5 - 1)(7 - 1) + 1 = 21$ [sic]. There is a mathematical relationship that says that $x = x^{21}$ *mod* 35 for any x from 0 to 34. Now you factor 21 yields 3 and 7. You pick one of those numbers to be your private key and the other one is your public key. So you have: Public key: 3 Private key: 7 Someone encrypts a message for you by taking plaintext message m to make cyphertext message c : $c = m^3$ *mod* 35. You decrypt c and find m using your private key: $m = c^7$ *mod* 35. If the numbers are several hundred digits long (as in PGP), it is nearly impossible to guess the secret key.'" (The calculation is actually incorrect: when I asked him, May explained that *Cyphernomicon* was only a first draft, and that he'd never got round to checking it as carefully as he would have liked.)

4 As explained in *Attack of the 50-Foot Blockchain* by David Gerard (CreateSpace, 2017), Szabo has studied law, and seems to take quite a cautious approach to this issue, unlike others.

5 Kelly Murnane, 'Ransomware as a Service Being Offered for $39 on the Dark Net', www.forbes.com, 15 July 2016.

6 See Gerard, *Attack of the 50-Foot Blockchain* for an excellent discussion of this issue.

7 Annie Nova, '"Wild west" days are over for cryptocurrencies, as IRS steps up enforcement', www.cnbc.com, 17 January 2018.

8 'A Simple Guide to Safely and Effectively Tumbling (Mixing) Bitcoin', https://darknetmarkets.org, 10 July 2015.

'Can the taxman identify owners of cryptocurrencies? www.nomoretax.eu, 7 September 2017.

9 IRS is going after Coinbase, ordering it last year to hand over details of 14,000 people who carried out big transactions; Robert Wood, 'Bitcoin Tax Troubles Get More Worrisome', www.forbes.com, 4 December 2017.

10 Amanda Taub 'How Stable Are Democracies? "Warning Signs Are Flashing Red"', www.nytimes.com, 29 November 2016.

There are lots of other similar surveys saying much the same thing: Yascha Mounk 'Yes, people really are turning away from democracy', www.washingtonpost.com, 8 December 2016; R. S. Foa and Y. Mounk (2016), 'The democratic disconnect', *Journal of Democracy*, 27 (3), 5–17.

Conclusion: Say Hello to the Future

1 Bruce Drake, '6 New Findings about Millennials', www.pewresearch.org, 7 March 2014.
2 David Runciman, 'How Democracy Ends' (a 2017 lecture).
3 Rachel Botsman, 'Big data meets Big Brother as China moves to rate its citizens', *Wired*, 21 October 2017.

Epilogue: 20 Ideas to Save Democracy

1 Goodman et al., 'The new political campaigning'.
2 Keen, *How to Fix The Future*, p.32.
3 Robert Mendick, 'Treasury crackdown on Bitcoin over concerns it is used to launder money and dodge tax', www.telegraph.co.uk, 3 December 2017.

Felicity Hannah, 'Bitcoin: Criminal, profitable, or bubble?', www.independent.co.uk, 11 December 2017.
4 For more on this, I recommend you read David Birch's excellent booklet, *Identity is the New Money*.

Acknowledgements

FIRST IN THE LONG line of thanks is the team at Ebury, whose professionalism, talent and belief in this project is the chief reason you're reading this book at all. My editor Andrew Goodfellow has improved the work in immeasurable ways, and Clare Bullock, Michelle Warner, Joanna Bennett, Clarissa Pabi and Caroline Butler have all been a joy to work with. Thanks also to the copy-editor, Nick Humphrey, and the proof reader, Katherine Ailes. As ever special thanks go to my agent Caroline Michel and other brilliant colleagues from PFD – how did I ever get anything done without them?

I've incurred further debts to all my Demos colleagues, who continue to persevere through my prolonged absences (especially Carl, Alex, and Josh). Friends, family or experts looked over and provided invaluable feedback: AKJ, Jon Birdwell, Tom Chatfield, Bob Greifinger, Alan Lockey, Polly Mackenzie, Martin Moore, Rick Muir, Simon Parker, Jack Rampling, Jeremy Reffin, Leo Sands, Thom Townsend and Alex Whitcroft. For any aspiring authors out there: share drafts often and early – it will improve your work immensely. I had a brilliant researcher whose work and dedication seemed to surpass even mine: Christopher Lambin. Big thanks, Chris. Thanks also to Alice Reffin, who was a brilliant help.

A handful of the interviews in this book were conducted as part of the BBC Two series *Secrets of Silicon Valley*. I'm grateful to those who agreed to take part, and the outstanding team I worked with, above all Ammar, Jack, Jamie (x2), Kerianna, Mike, Sam, Seb and Tristan.

Finally to Catrin, without whom the project would still be thrashing around in the back of my mind somewhere, and who supported me every moment, from start to finish.